Springer Series in Advanced Manufacturing

Series Editor

Duc Truong Pham, University of Birmingham, Birmingham, UK

The **Springer Series in Advanced Manufacturing** includes advanced textbooks, research monographs, edited works and conference proceedings covering all major subjects in the field of advanced manufacturing.
The following is a non-exclusive list of subjects relevant to the series:

1. Manufacturing processes and operations (material processing; assembly; test and inspection; packaging and shipping).
2. Manufacturing product and process design (product design; product data management; product development; manufacturing system planning).
3. Enterprise management (product life cycle management; production planning and control; quality management).

Emphasis will be placed on novel material of topical interest (for example, books on smart manufacturing and sustainable manufacturing) as well as new treatments of more traditional areas.

As advanced manufacturing usually involves extensive use of information and communication technology (ICT), books dealing with advanced ICT tools for advanced manufacturing are also of interest to the Series.

Springer and Professor Pham welcome book ideas from authors. Potential authors who wish to submit a book proposal should contact Anthony Doyle, Executive Editor, Springer, e-mail: anthony.doyle@springer.com.

INDEXED IN SCOPUS

Sihan Huang · Xingyu Li · Xi Gu · Yoram Koren
Editors

Design and Operation of Smart Reconfigurable Manufacturing Systems in Industry 4.0/5.0

Springer

Editors
Sihan Huang
School of Mechanical Engineering
Beijing Institute of Technology
Beijing, China

Xingyu Li
School of Engineering Technology
Purdue University
West Lafayette, IN, USA

Xi Gu
Department of Mechanical and Aerospace Engineering
The State University of New Jersey
Piscataway, NJ, USA

Yoram Koren
Department of Mechanical Engineering
The University of Michigan
Ann Arbor, MI, USA

ISSN 1860-5168 ISSN 2196-1735 (electronic)
Springer Series in Advanced Manufacturing
ISBN 978-3-032-00283-9 ISBN 978-3-032-00284-6 (eBook)
https://doi.org/10.1007/978-3-032-00284-6

This Springer imprint is published by the registered company Springer Nature Switzerland AG
The registered company address is: Gewerbestrasse 11, 6330 Cham, Switzerland

Preface I

The globalization that emerged in the early 1990s made the "rapid responsiveness" to fast changing markets a vital goal for industry. Traditional dedicated production lines, which dominated the manufacturing of high-precision parts (e.g., car engines and transmissions in the automotive industry) at that time, could not meet the demand goal. Therefore, Dr. Yoram Koren proposed the "Reconfigurable Manufacturing System (RMS)" as a novel, next-generation manufacturing system in his 1995 Engineering Research Center (ERC) proposal to the National Science Foundation (NSF). The proposal was awarded, and the ERC-RMS was established on August 1, 1996. From 1996 to 2007, the ERC-RMS received $32.5 million from the NSF to develop the RMS scientific basis, technologies, and software and hardware tools as well as 16 U.S. patents. The ERC-RMS was the first ERC that the University of Michigan won in its history, and it was formally recognized as a major milestone in the U-M College of Engineering 150-year history.

During the last two decades, RMS has become a worldwide research discipline with practical implementations in factories. Today, most powertrain factories (that produce car engines and transmissions) in the U.S. are designed according to the RMS principles and the RMS system-architecture, and possess the RMS characteristics. Furthermore, many of the RMS technologies have become enabling technologies that are essential to improve the manufacturing flexibility and intelligence in today's Industry 4.0/5.0 era.

Design and Operation of Smart Reconfigurable Manufacturing Systems in Industry 4.0/5.0 is a timely book that was initiated by Prof. Sihan Huang. It includes significant chapters that elaborate on important aspects of the design and operation of RMS in the current era, and how the synthesis between RMS technologies and the other Industry 4.0/5.0 technologies can shape the smart factories of today and the future.

We hope that you will enjoy this book.

Prof. Yoram Koren
James J. Duderstadt Distinguished
University Professor Emeritus
of Manufacturing
Paul G. Goebel Professor Emeritus
of Engineering
The University of Michigan
Ann Arbor, USA

Preface II

During the last decade, Industry 4.0 as the symbol of a new industrial revolution has rapidly gained popularity around the globe. Industry 4.0 technologies—primarily IT-driven—have been increasingly implemented in today's factories and improve the production efficiency and quality to a higher level. Industry 5.0 is a forethoughtful concept on the future of industry towards a human-centric, sustainable, and resilient manufacturing system (Huang et al. 2022), which provides a flourishing development vision of industry. The parallel development of Industry 4.0 and Industry 5.0 produced lots of interesting research and applications.

In the Industry 4.0/5.0 era, the manufacturing paradigm is shifting from mass customization to mass individualization/personalization, which requires more flexible and efficient manufacturing systems that are equipped with reconfigurable facilities and driven by smart technologies. A Reconfigurable Manufacturing System (RMS), invented by Prof. Koren in the 1990s, provides the exact capacity and functionality needed, exactly when needed. It responds quickly and cost-effectively to demand fluctuation through multi-level reconfiguration (e.g. system level, machine level). Since its invention, RMS has received worldwide interest from academia and industries, with research topics on various aspects including RMS design, operations, industrial implementations, etc. Smart manufacturing is a core technology to handle demand fluctuation through self-perception, self-learning, self-decision-making, self-execution, and self-adaptation, etc. A smart RMS (named SRMS) could promote the ability of handling uncertain demand more effectively and efficiently.

This book presents the evolution roadmap of RMS and its characteristics since its invention. It discusses the role of RMS in today's increasingly competitive manufacturing environment and unpredictable markets, and the new challenges and opportunities in the RMS design and operations in the emerging Industry 4.0/5.0 era. In addition, this book investigates the relationship between reconfigurable manufacturing and smart manufacturing and demonstrates how smart manufacturing techniques can facilitate the design and operations of RMSs. This book investigates the design and operations of SRMS in the Industry 4.0/5.0 era for addressing the recent emerging challenges: (1) The influence of new demand characteristics in Industry 4.0/5.0 on RMS, (2) How to integrate the new information and communication technologies

(ICT) and artificial intelligence (AI) with the design and operation of RMS to evolve RMS to SRMS, (3) New methodologies on the design and operations of SRMS, (4) Optimization of SRMS by considering the features of Industry 4.0/5.0.

Ten chapters were selected after a peer review process. They were revised in accordance with the suggestions and recommendations from the reviewers.

Chapter "The Roadmap of Reconfigurable Manufacturing System: From Its Emergence to the Era of Industry 4.0 and 5.0" by Xi Gu et al., presents the roadmap of the RMS paradigm by reviewing its emergence and its development trend. It also discusses the role of RMS in the current In-dustry 4.0 and 5.0 era, and how the integration of RMS technologies and other emerging technologies can improve the competitiveness of today's manufacturing enterprises.

Chapter "Intelligent Clustering Method of Part Family Formation for D-RMS" by Sihan Huang et al., addresses part family formation problem of delayed reconfigurable manufacturing systems (D-RMS). Firstly, a exclusive similarity coefficient of D-RMS is constructed. Secondly, a machine learning method named K-Medoids is adopted to cluster parts into families based on similarity. Finally, a case study is conducted to demonstrate the proposed method.

Chapter "A NSGA-II-Based Optimization Method for Configuration Design of D-RMS" by Sihan Huang et al., proposes a optimization method for D-RMS configuration design based on NSGA-II. First, a multi-objective optimization model is formulated, incorporating factors such as machine tool investment costs, reconfiguration costs, reconfiguration time, and delayed reconfiguration. Second, NSGA-II is employed to solve the proposed optimization model. Finally, a case study is conducted to validate the effect of the proposed method.

Chapter "Online Optimization of Reconfiguration Planning for SRMS Based on DQN" by Sihan Huang et al., investigates a reconfiguration planning optimization method for smart reconfigurable manufacturing systems (SRMS) based on deep Q-network (DQN). The reconfiguration processes of smart manufacturing systems are modeled by considering reconfiguration cost, moving cost, and processing cost. DQN is adopted to explore the state space and action space to seek optimal reconfiguration policy online.

Chapter "Enhancing Resilience in Reconfigurable Manufacturing Systems Through Integrated Production and Maintenance Planning" by Xingyu Li et al., proposes a degradation-aware RMS decision-making model to optimally determine and adjust operational actions in real-time considering demand fulfillment, maintenance cost, and system health. The proposed approach has the capability to capture the causality between operational action sequences and the resulting system deterioration through smart manufacturing technologies and artificial intelligence-based methods.

Chapter "Autonomous Coordination of Multiple Reconfigurable Manufacturing Systems" by Xingyu Li et al., investigates a manufacturing environment comprising multiple reconfigurable production lines that are designed to simultaneously produce a diverse product mix. These lines share modular components, enabling dynamic reconfiguration of each line in real time to adapt both capacity and functionality. In response to the advancements brought by smart manufacturing, we propose a real-time teaming of reconfigurable manufacturing systems characterized by decentralized management across autonomous production lines and coordinate the multiple lines to collectively satisfy the market changes.

Chapter "A Multi-Agent Deep Reinforcement Learning for Flexible Job Shop Scheduling Problem with Transportation" by QiXin Wang et al., proposes a multi-agent Deep Reinforcement Learning (DRL) framework to address the Flexible Job Shop Scheduling Problem with Transportation (FJSPT). Experimental results demonstrate that, after individual training, the two agents in the proposed framework outperform handcrafted heuristic dispatching rules in terms of solution quality.

Chapter "Industry 4.0 Hyper-Factory Planning, Scheduling, and Execution with Real-Time Data" by Mingxing Li et al., proposes a novel Industry 4.0 PSE methodology, Hyper-Factory Planning, Scheduling, and Execution (HF-PSE), designed to enable adaptive and robust decision-making amidst diverse uncertainties in contemporary manufacturing settings. The case study demonstrates that HF-PSE surpasses others on average and exhibits minimal statistical variations in cost-efficiency, punctuality, and simultaneity metrics, highlighting its superior effectiveness, stability, and resilience in stochastic settings.

Chapter "Reconfigurable Manufacturing Systems for Industry 4.0: Concepts, Challenges, and Applications" by Muhammad Ameer and Mohammed Dahane, provides an overview of how manufacturing systems have progressively evolved toward RMS within the context of Industry 4.0, which focuses on the key challenges in RMS design, the selection of reconfigurable system components, and process planning strategies in RMS environments.

Chapter "Reconfigurable Supply Chain Network Design: Leveraging RMS Insights" by Shokraneh K. Moghaddam and Angela D. Traill, presents a comparative analysis of RMS and Reconfigurable Supply Chains (RSC), examining their shared principles, distinct features, and the potential to transfer RMS insights to RSC development. The shift from RMS to RSC involves unique challenges including the complexity of coordinating multiple stakeholders, ensuring data interoperability, achieving end- to-end visibility, and balancing reconfiguration costs with performance benefits.

We hope that you will enjoy reading this book.

Beijing, China — Sihan Huang
West Lafayette, USA — Xingyu Li
Piscataway, USA — Xi Gu
Ann Arbor, USA — Yoram Koren

Acknowledgements

We sincerely express our heartfelt gratitude to Series Editor Prof. Duc Truong Pham and Executive Editor Dr. Anthony Doyle for their invaluable guidance and support throughout the book editing process. We also deeply appreciate the Springer editorial team for their active involvement and support, whose contributions have been instrumental in the publication of this book.

We extend our sincere appreciation to all reviewers for their insightful comments and constructive feedback, and to the authors for their impressive manuscripts. Additionally, Sihan Huang acknowledges the funding support of the National Natural Science Foundation of China (52405530), Xingyu Li acknowledges the funding support of National Science Foundation (2531898). Without the invaluable assistance and support, the publication of this book would not have been possible.

Special thanks to Ming Huang, Zhicheng Peng, Guangyu Mo, Jianpeng Chen, Jintang Huang, and Qizhang Zhu for their assistance to this book.

Contents

The Roadmap of Reconfigurable Manufacturing System: From Its Emergence to the Era of Industry 4.0 and 5.0

Xi Gu, Yoram Koren, Xingyu Li, and Sihan Huang

Abstract The term "Reconfigurable Manufacturing System (RMS)" was coined by Prof. Koren three decades ago, in response to the challenges caused by global manufacturing. RMS has rapid responsiveness and can provide the exact capacity and functionality needed, exactly when needed. Since then, many RMS models, principles, and technologies, ranging from machine level to system level and supply-chain level, have been invented. RMS has been established as a new scientific research area that has gained increasing attention and interest from both academia and industry. In the current Industry 4.0 and 5.0 era, new technologies (e.g., cyber-physical systems, digital twins, artificial intelligence, data analytics) and new manufacturing goals (e.g., market-of-one products, sustainability, resilience, human-centricity) are emerging. This chapter presents the roadmap of the RMS paradigm by reviewing its emergence and its development trend. It also discusses the role of RMS in the current Industry 4.0 and 5.0 era, and how the integration of RMS technologies and other emerging technologies can improve the competitiveness of today's manufacturing enterprises.

Keywords Reconfigurable manufacturing system · Manufacturing paradigm · Industry 4.0 · Industry 5.0

X. Gu (✉)
Department of Mechanical and Aerospace Engineering, Rutgers, The State University of New Jersey, Piscataway, NJ, USA
e-mail: xi.gu@rutgers.edu

Y. Koren
Department of Mechanical Engineering, The University of Michigan, Ann Arbor, MI, USA

X. Li
School of Engineering Technology, Purdue University, West Lafayette, IN, USA

S. Huang
School of Mechanical Engineering, Beijing Institute of Technology, Beijing, China

S. Huang et al. (eds.), *Design and Operation of Smart Reconfigurable Manufacturing Systems in Industry 4.0/5.0*, Springer Series in Advanced Manufacturing,
https://doi.org/10.1007/978-3-032-00284-6_1

1 Introduction

A manufacturing paradigm is characterized by its goal and innovations in the product, system, and business model (and their integrations), in response to the contemporary societal needs and market competition. The RMS paradigm was initiated in the late 1990s, when many manufacturing companies started to compete in a global market and build their factories overseas. On a global market, the customer demands were unpredictable, and the manufacturing systems should be responsive to meet such abruptly changing demand quickly and cost-effectively. To address these challenges, Prof. Koren coined the term "Reconfigurable Manufacturing System (RMS)" in his 1995 Engineering Research Center (ERC) proposal to the US National Science Foundation (NSF). The goal of the RMS is to provide the exact production capacity and functionality needed, exactly when needed by the market, through rapid and cost-effective reconfiguration of the manufacturing resources.

The NSF awarded Koren's proposal, and on August 1, 1996, the ERC for RMS (ERC-RMS) was established at the University of Michigan. The ERC-RMS created the scientific basis of the RMS paradigm, and developed software packages and hardware machines (e.g., reconfigurable machine tool and inspection station). The ERC-RMS received a total funding of $47 million from 1996 to 2012, of which 32.5 million was from a National Science Foundation grant during 1996–2007, and the rest was from industry and the state of Michigan. The ERC-RMS has also created the largest academic manufacturing R&D testbed in the US. During the past 30 years, RMS has become a worldwide research discipline with many successful technological innovations and practical implementations in factories around the world.

New technologies and societal needs are continuously shaping the manufacturing paradigm. Currently, we are experiencing the "Fourth Industry Revolution" (i.e., Industry 4.0), enabled by the rapid development of new manufacturing and automation technologies (e.g., additive manufacturing, robotics) as well as information and communication technologies (e.g., Internet-of-Things, Digital Twin, big data). These technologies are increasingly being implemented in today's factories and improving their manufacturing performance. In addition, there has been a recent trend that proposes to shift the primary goals on the increasing production efficiency to addressing other social needs, such as promoting the well-being of workers and promoting the sustainability, which is called "Industry 5.0". "Industry 4.0" and "Industry 5.0" address various aspects of the current era. In the current era of Industry 4.0 and 5.0, many new manufacturing concepts and models are developed, and many of them have RMS technologies as technological enablers or integrative elements.

This chapter presents the roadmap of RMS, including its emergence and development during the past three decades. We review the key RMS characteristics, principles, inventions, and research topics. We will also discuss the role of RMS in the current Industry 4.0 and 5.0 era, and how the integration of RMS and other emerging technologies is shaping the factory of today and the future.

2 The Emergence of the Reconfigurable Manufacturing Systems Paradigm

The decade from 1991 to 2001 witnessed intensified globalization, due to (1) the changes in governmental policies (e.g., the creation of the EU and NAFTA, India's and China's joining WTO), (2) the development of worldwide transoceanic fiber-optic cable networks that increased the amount of information flow, as well as (3) the global expansion of manufacturing industry (Koren 2010). The huge number of competitors and customers in a global market raised significant challenges for manufacturing enterprises to stay competitive. Because the customers were from different countries and cultures, product variants needed to be developed to meet the requirements of diverse customers; because of the increased competition, new products would be developed and introduced to the market at a faster pace, resulting in shorter product lifecycles. These factors, coupled with the dramatically complex dynamics of the global supply chain, made it extremely difficult to accurately predict the customer demand (in terms of the product volume and functionality). To improve the competitiveness in the global market, manufacturing enterprises need to develop strategies to respond rapidly and cost-effectively to market changes, and to produce high-quality products.

During the second half of the twentieth century, dedicated manufacturing lines (DMLs) dominated the manufacturing of high-precision parts in the automotive industry worldwide. Although these dedicated lines could accomplish the goals of high-quality and low-cost products, they were based on fixed automation and were typically used to produce the same parts in high volumes. This is suitable in the ideal case in which the demand is predictable and is constant. However, the uncertainties raised in the global market made DMLs not an optimal economical solution: If the demand volume would increase or decrease in the future, the system with DMLs would either be underproduce (loss of revenue) or idle (waste of capacity). In addition, since the DMLs are focused on a single product and cannot handle product variants, when new product generations and variants are introduced, new lines would be built with old lines discarded. Another type of the existing manufacturing system at that time was flexible manufacturing systems (FMSs) that consist of computer numerically controlled (CNC) machines. However, FMSs typically use general purpose technology to produce customized products. They were slow and could not increase the production rate to match increased market demand, so the production was not cost-effective. Therefore, both DMLs and FMSs could not sufficiently handle the challenges brought by globalization, and developing a new type of manufacturing system became critical.

The key challenge is to develop a manufacturing system for the right mix of products with the technical capabilities that allow for the quick changeover of product mix quantities that might vary dramatically. RMS was proposed as a solution to quickly respond to such a challenge. It combines the advantages of both DMLs (i.e., low cost and high product quality) and FMSs (i.e., the capability to produce various products) by focusing the level of flexibility/customization around a part family.

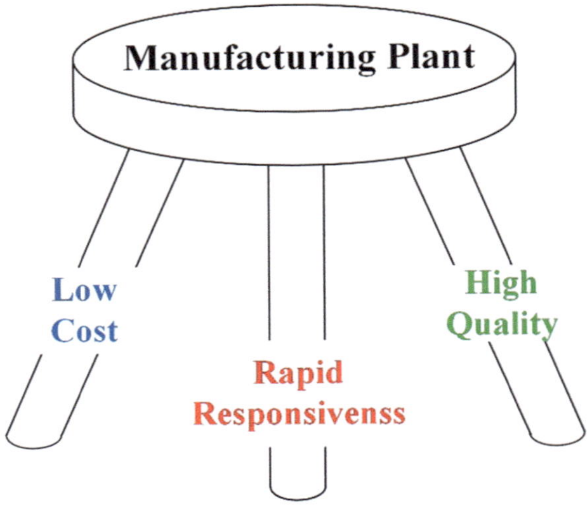

Fig. 1 The three goals for a modern manufacturing plant: low cost, high quality, and rapid responsiveness

RMS can respond quickly and cost-effectively to the market changes by providing "exactly the capacity and functionality needed, exactly when needed" (Koren et al. 1999).

The rapid responsiveness brought by RMS is one of the key goals for modern manufacturing plants. A modern manufacturing plant for high-volume production has three goals (as shown in Fig. 1, all the three legs for the stool need to be strong): 1. **Low-cost** that was achieved by the moving assembly line that enables mass-production factory, which was invented by Henry Ford in 1913, to make owing a car affordable. 2. **High-quality** that was achieved by John Parsons who invented the first numerical control (NC) machine in the 1940s. 3. **Rapid Responsiveness** of the factory production-rate to unexpected changes in the market demand of a certain product, which was invented by Yoram Koren in 1998 (Koren's RMS Patent: US 6,349,237). The rapid responsiveness of the manufacturing system enables profit to the producers even in the global fluctuating markets. Today most engine and transmission manufacturing factories of Ford, GM, and Chrysler are designed with the RMS system architecture.

3 RMS Scientific Basis and Research Topics

Since the RMS architecture and its scientific basis were created, RMS has received increasing interest from both academia and industry. RMS opened new research areas in the design and operations of various types of reconfigurable machines and reconfigurable manufacturing systems. In particular, researchers in the ERC/RMS have done a lot of pioneering research and established the RMS scientific basis. In this section, we will first review some of the work done at the ERC-RMS, followed by a survey on the RMS research topics in the literature.

3.1 The RMS Scientific Basis Developed at the ERC-RMS

The early research developed at ERC-RMS has defined the RMS characteristics and design principles, which have laid the foundation for the RMS research. The core RMS characteristics include customization, modularity, integrability, scalability, convertibility, and diagnosability:

1. **Customization** means RMS is designed around a customized flexibility (i.e., a product family) that combines the advantages of the cost-effectiveness of DMLs and the flexibility of FMSs.
2. **Modularity** means RMS is designed based on a modular structure. At the system level, the machines and materials handling equipment (gantries, conveyors, automated guided vehicles, etc.) are modules that can be easily added or replaced; while the various components (axes, tooling, software, etc.) of the machines and equipment may also be modules. Such a modular design facilitates the rapid and cost-effective reconfiguration.
3. **Integrability** means the integration of different modules (at the machine level, system level, and control level) to form the reconfigured machine and system.
4. **Scalability** means the capability for the system to change its capacity quickly and cost-effectively in response to the volume changes on the market. A mathematical definition of scalability can be found in Wang and Koren (2012).
5. **Convertibility** means the capability for the system to change its functionality quickly and cost-effectively in response to the new product requirement. A mathematical definition of convertibility can be found in Maler-Speredelozzi et al. (2003).
6. **Diagnosability** means the ability to detect the part defects and identify the root-cause machines when the part defects occur, so that the part quality can be improved. In multi-stage RMS that has multiple stages and parallel machines, many product streams exist. "Stream-of-variations" (SoV) methods (Shi 2006) are developed to analyze how the uncertainties in each machine and process can attribute to the quality variation of the final products. In addition, reconfigurable inspection machines can be embedded in the system for online quality detection. These tools enable RMS to start producing higher quality parts quickly after reconfiguration.

These six core characteristics provide guidance on the design and operations of RMS and reconfigurable machines. Based on these core characteristics, the principles for designing RMSs can be developed as follows (Koren and Shpitalni 2010).

- Designing the system and its machines for adjustable structure that facilitates system scalability in response to market demands and system/machine convertibility to new products. Structure may be adjusted (1) at the system level (e.g., adding machines), (2) at the machine level (e.g., adding spindles and axes, or changing angles between axes), and (3) at the control software (e.g., integrating easily advanced controllers).
- Designing the manufacturing system around the part family, with the customized flexibility required for producing all parts of this part family.

In addition to the new system architecture and the scientific basis, original prototypes of reconfigurable machines and RMS have been built. The researchers at ERC-RMS have obtained twelve US patents on RMS—at both the machine-level and the system-level—that legally established the RMS invention. These patents include:

- Reconfigurable Machine Tool (US 5,943,750) by Y. Koren and S. Kota
- Reconfigurable Manufacturing System (US 6,349,237) by Y. Koren and G. Ulsoy
- Reconfigurable Logic Controller (US 6,256,598) by E. Park, D. Tilbury, and P. Khargonekar
- Reconfigurable Power Spindle (US 6,309,319) by Y. Moon and S. Kota
- Reconfigurable Automatic Tool Changer (US 6,442,815) by Y. Moon and S. Kota
- Reconfigurable Apparatus and Method for Inspection (US 6,567,162) by Y. Koren and R. Katz
- Bi-axial coplanar apparatus (US 6,567,235) by R. Katz, Y. Koren, and F. Pierrot
- Reconfigurable Multi-spindle Apparatus (US 6,569,071) by Y. Koren. Y. Moon, and S. Kota
- Optical Method for Measuring angular alignment (US 6,674,521) by S. Segall and J. Upatnieks
- Integrated Reconfigurable Manufacturing System (US 6,920,973) Y. Koren and R. Hill
- Reconfigurable Surface Finish Inspection Apparatus for cylinder bores (US 7,027,145) by S. Segall and Y. Koren
- Reconfigurable Linescan Illumination (US 7,168,822) by G. Abramovich, Z. Warlick, and Y. Koren.

These inventions enable the RMS core characteristics and design principles to be implemented in practice. A recent survey on these patents (Koren 2022) shows that three of them have the biggest impact (i.e., the highest citations by other patents and papers): Reconfigurable Manufacturing System, Reconfigurable Machine Tool (RMT), and Reconfigurable Inspection Machine (RIM). These inventions are critical elements of the RMS. Figure 2 shows an RMS configuration with CNCs, RMTs and RIMs. Such an RMS configuration has a higher scalability and convertibility than DMLs, and a Return System is embedded to transfer the defective parts for rework. This system architecture can also improve the system flexibility to effectively produce mass-individualized products. Additional spaces are reserved in the system so incremental capacity can be quickly added in the future, when needed.

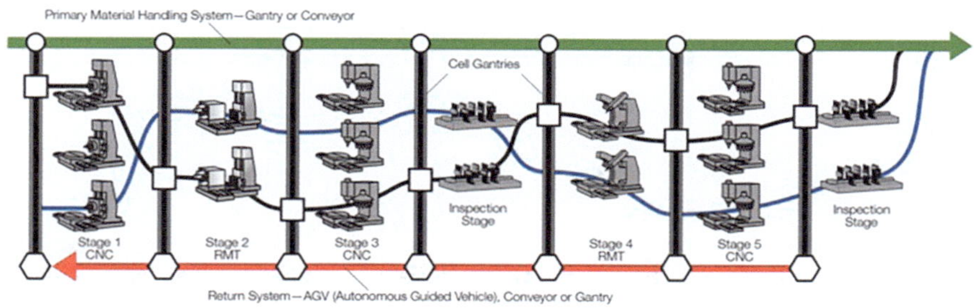

Fig. 2 Reconfigurable manufacturing system and machines

The ERC-RMS had three research thrust areas: (1) system-level design, (2) machine and control design, and (3) ramp-up and operations. Many RMS research problems in these three thrust areas were investigated, and consequently, many RMS technologies—including theories, methodologies, and testbeds—were developed. Some of these RMS research problems and technologies include:

- **System-level design**: system configuration analysis, life-cycle economic modeling, system-level process planning
- **Machine and control design**: reconfigurable machine design theory, open-architecture controller design
- **In-process metrology and operations**: Stream-of-Variations (SoV) theory, and embedding reconfigurable inspection machines in the manufacturing system

Figure 3 shows the relations between these RMS technologies and the corresponding RMS core characteristics addressed. Technologies in different thrust areas are represented by different colors.

The ERC-RMS has developed several original software packages based on the theories and methodologies, including: (1) Life-Cycle Economic Modeling software package, (2) Performance Analysis of Manufacturing Systems (PAMS), (3) Software for Handling the Assignment of tasks to Resources (SHARE), which focuses online balancing and RMS design and improvement. These software packages and other RMS technologies have been successfully implemented in industry. Some technology transfer activities at the Big Three automakers are summarized in Table 1.

These RMS characteristics, principles, inventions, patents, and technologies developed at the ERC-RMS have established the scientific basis of RMS, especially for the automotive industry. Furthermore, many new RMS theories, methodologies, and prototypes of machines and systems, have been developed by researchers and industrial practitioners around the world. The RMS technologies have also been integrated with other technologies and extended to various industries and aspects to address the related problems faced by global manufacturing enterprises.

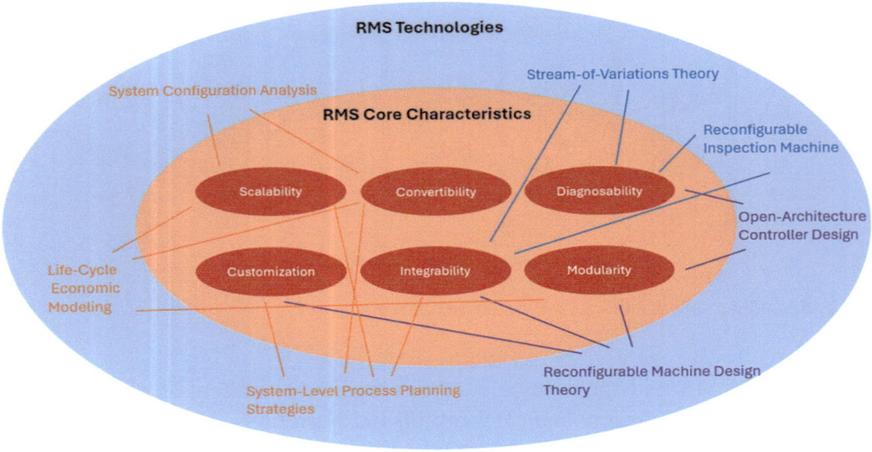

Fig. 3 RMS core characteristics and technologies

3.2 A Survey on RMS Research Topics and Trend

To understand the RMS research topics and trend in literature, we did a survey on the RMS literature until 2024. Our search utilized the query "*reconfigurable manufacturing/manufacture*" (Topic) OR "*reconfigurable machine*" (Topic), yielding approximately 4,900 relevant articles. As illustrated in Fig. 4, the increasing number of publications in reconfigurable manufacturing indicates an increase interest in this area over the years.

Among these publications, many review papers have been published to summarize and analyze RMS research from various aspects. Some of the review papers are given below to provide an overview of the RMS research topics and problems over the time.

Regarding the design of Reconfigurable Machines:

- Bi et al. (2008b) reviewed reconfigurable machines developed in reconfigurable machining systems (e.g., RMTs, reconfigurable dies, parallel kinematic machines, reconfigurable fixturing systems, reconfigurable assembling systems, and reconfigurable material handling systems). The design issues and methodologies of these reconfigurable machines were also discussed.
- Gadalla and Xue (2017) reviewed the recent advances in the development of RMTs to overcome the challenges in three aspects: (1) architecture design, (2) configuration design and optimization, and (3) system integration and control. It classified the architecture of RMTs into modular architecture (which can be further classified into open-architecture and semi-open architecture) and integral architecture.
- Morgan et al. (2021) explored the state-of-the-art in distributed and decentralized machine control and machine intelligence and established a vision

Table 1 Technology transfer at the "Big Three"

	Chrysler	Ford	General motors
Manufacturing system design	• Application of PAMS to improve productivity—Indiana Transmission Plant • Automatic Line Balancing (SHARE) Applied to Assembly Lines • Life Cycle Economic Modeling	• Application of PAMS to improve productivity at Cleveland Engine Plant • Simulation Models and Validation for Complex Manufacturing Systems • Life Cycle Economic Modeling	• Application of PAMS and Line Balancing software into OPTline • Performance Analysis of Parallel and Serial Machining Lines • Process Planning Powertrain Protocol
Machines and control	• Open Factory-wide Control Architecture for reduction of unscheduled downtime • Networked Control Testing Methodology	• Evaluation and Comparison of Automatic Logic Control Generation Approaches • Study of Reduced Tool Life in Plunging Operations	• Evaluation of EtherNet IP and Networked Control • Real-Virtual Integration of Control Systems and Networks
In-process metrology and system operations	• Porosity Inspection at GEMA Plant	• Plan for Porosity Inspection at Lima. Ohio Plant	• Porosity Inspection at Flint Engine Plant
	• Cylinder Bore Surface Finish Inspection • Optimization of measurements intervals for Reduction of Measurement Data	• Supports of STTP on Cylinder Bore Surface Inspection • (SoV) Application and Data-driven Measurement Reduction Windsor Engine Plant	• Interferometric based Cylinder Bore Surface Inspection • Data-driven Measurement Reduction (reduce CMM time)

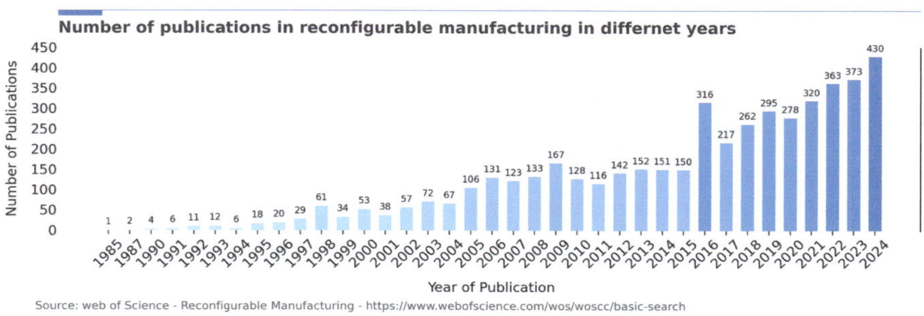

Fig. 4 Trends in annual publications on reconfigurable manufacturing

for next-generation Industry 4.0 manufacturing machines, which would exhibit extraordinary smart and reconfigurable capabilities.

Regarding the design and operations of RMS:

- Bi et al. (2008a) provided a state-of-the-art review of on RMS, and discussed the strategies to meet customer requirements such as short lead-time, more variants, fluctuating volumes, and low cost, and classified the RMS design issues into architecture design (i.e., determine the system components and their interactions at the phase of system design), configuration design (i.e., determine the system configuration given the architecture and specific task at the phase of system application), and control design (i.e., determine the process variables at the phase of system operations) and discussed their corresponding methodologies.
- Renzi et al. (2014) reviewed the roles of meta-heuristic methods (e.g., tabu search, simulated annealing, particle swarm optimization, genetic algorithm, and ant colony optimization) and artificial intelligent methods (e.g., artificial neural networks, fuzzy logic) in solving RMS design problems such as generation/selection of the optimal cell configuration and process-planning.
- Andersen et al. (2017) classified the design methods for RMS in literature into phased design method and cyclic design method and did comparative analysis and synthesis towards a generic design method.
- Bortolini et al. (2018) classified the reviewed papers on RMS into five streams: (1) reconfigurability level assessment, (2) analysis of RMS features (i.e., the six core characteristics), (3) analysis of RMS performance (e.g., quality, availability, system complexity, responsiveness, sustainability), (4) applied research and field applications (e.g., product family formation, configuration, layout, planning and scheduling), and (5) reconfigurability towards Industry 4.0 goals.
- Yelles-Chaouche et al. (2021) reviewed the RMS problems from an optimization perspective. It discussed different types of reconfigurable machines (i.e., reconfigurable fixtures, reconfigurable machine tools, reconfigurable assembly machines, and reconfigurable inspection machine) and RMSs (reconfigurable flow lines, reconfigurable rotary machining systems, and reconfigurable cellular manufacturing system), classification of the optimization problems (including RMS design, production planning and scheduling, layout design, line balance and re-balancing) as well as their objective functions and solution approaches.

In addition to the automotive industry, an increasing amount of literature focuses on designing RMS for other industry applications. For example, Arista et al. (2022) did a systematic literature review on RMS design for aerospace industry, particularly, on the role of the Industrial Resources in during the RMS conceptual design phase. It identified and analyzed three streams of literature: (1) Industrial Resources characteristics at conceptual design phase, (2) decisions involving Industrial Resources in RMS design, and (3) Industrial Resources relationship to Product and Processes in RMS design.

Notably, some of the recent literature reviews on RMS indicate synthesis between RMS and Industry 4.0 or smart manufacturing, e.g., Morgan et al. (2021) for "Industry

4.0 smart reconfigurable manufacturing machines" and Bortolini et al. (2018) for "reconfigurability towards Industry 4.0 goals". Li et al. (2024) did a literature review on the relationship between reconfigurable manufacturing and smart manufacturing and found that there is a converging trend between the literature on these topics. This converging trend indicates the correlation between reconfigurable manufacturing and smart manufacturing regarding their technologies and goals, i.e., a more reconfigurable system is smarter, and a smarter system can be more reconfigurable.

Figure 4 also shows that the number of annual publications on RMS increases dramatically after 2016, suggesting a growing academic interest in the field, likely driven by the emergence of Industry 4.0 and 5.0 initiatives and the technological advancements aimed at enhancing the efficiency, flexibility, and smartness of manufacturing systems. To understand the impact of the Industry 4.0 on the RMS, we conducted a comparative analysis of abstracts from existing studies in RMS, distinguishing between those published before and after 2010, around what time when Industry 4.0 was first introduced. To process the abstracts, we applied lemmatization and tokenization to construct a unigram corpus, filtering out words with a total frequency of less than 100. For improved visualization, we incorporated semantic meanings, corpus frequencies, and cosine similarity to manipulate the positioning and coloring of points, each representing a high-frequency word, to enhance the clarity and interpretation of the distinct thematic areas. Figure 5 presents a visual representation of the distribution of key terms associated with RMS before and after 2010, alongside their corresponding word frequencies.

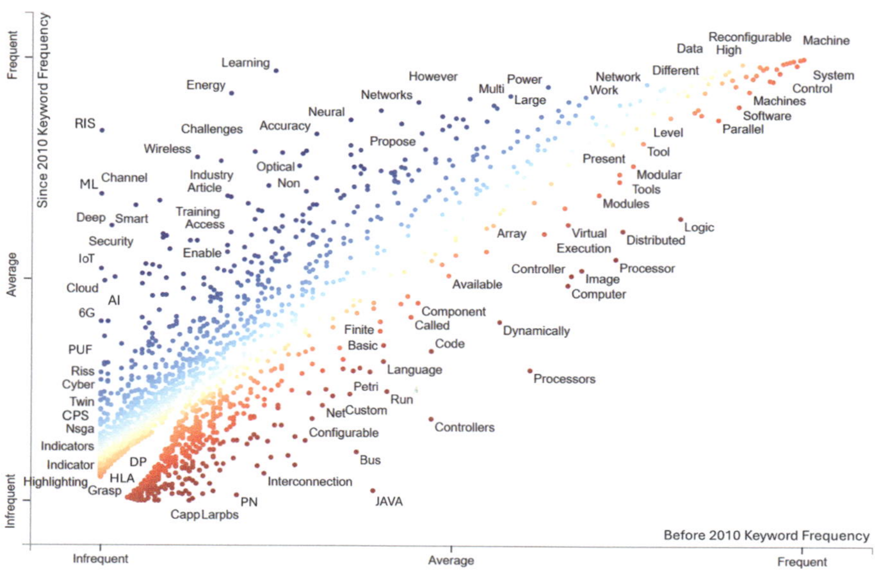

Fig. 5 Comparison of word distribution for RMS literatures before 2010 and after 2010

For the keyword across both periods (i.e., before and after 2010), we have *Machine*, *System*, *Reconfigurable*, *Control*, *Software*, and *Parallel*. Specifically, *Machine*, *System*, and *Parallel* emphasize the physical design of reconfigurable machines and system architectures, enabling flexibility in production capabilities. Meanwhile, *Control* and *Software* highlight the increasing role of software-driven reconfiguration, where advanced control strategies and parallel computing facilitate real-time adjustments and system optimization. These terms indicate a consistent focus on foundational principles in RMS.

The early research on RMS primarily focused on the enabling technologies necessary to achieve manufacturing reconfigurability. This emphasis is reflected in the frequent occurrence of terms such as *Processors*, *Logic*, *Controllers*, and *Computer*, which highlight the foundational role of computational hardware and control systems in facilitating adaptability within manufacturing environments. Additionally, terms like *Modular*, *Tools*, *Modules*, *Bus*, and *Interconnections* underscore the significant efforts devoted to enhancing the modularity of manufacturing resources and improving system interoperability. These keywords suggest a strong focus on developing modular hardware and communication architectures that enable flexible system configurations, facilitating rapid reconfiguration and adaptation to varying production requirements. Furthermore, other terms such as *PN* (*Petri Net*) and *DP* (*Dynamic Programming*) indicate the prevalent use of traditional mathematical modeling and optimization techniques for decision-making in reconfigurable manufacturing.

In contrast, with the rise of Industry 4.0, an increasing number of advanced technologies have been integrated into reconfigurable manufacturing, particularly emphasizing digital connectivity and intelligent automation. This trend is reflected in the prevalence of terms such as *IoT*, *Neural Network*, *Cloud*, *CPS*, *Digital Twin*, *Wireless*, and *Security*, which highlight the growing focus on leveraging information and communication technologies to enhance the interoperability, efficiency, and security of reconfigurable manufacturing systems. These advancements signify a paradigm shift toward the seamless integration of cyber-physical systems, ensuring both real-time responsiveness and secure data exchange within reconfigurable manufacturing environments. Moreover, the emergence of terms like *Data*, *AI* (*Artificial Intelligence*), *ML* (*Machine Learning*), *Training*, and *Deep Learning* underscores the transition from traditional mathematical programming approaches to data-driven methodologies in managing RMSs, reflecting the increasing reliance on artificial intelligence and machine learning for optimizing decision-making following the advent of industry 4.0.

The literature survey demonstrates an increasing amount of literature on the analysis of RMS in the context of Industry 4.0 and 5.0. In the next section, we will discuss the role of RMS in the current Industry 4.0 and 5.0 era, and how RMS and Industry 4.0 and 5.0 affect each other.

4 RMS Towards Industry 4.0 and 5.0

4.1 The Current Era of Industry 4.0 and 5.0

The landscape of manufacturing is continuously changing, driven by the customer demand and new technological advancements. Particularly, the rapid development of advanced manufacturing technology (such as digital manufacturing) as well as information and communication technology (such as AI) is significantly changing today's manufacturing systems. Currently, it is widely believed that we are in the era of Industry 4.0, or some researchers believe now is already in the Industry 5.0 era.

The term "Industry 4.0" (derived from the German term Industrie 4.0) was publicly introduced in 2011 in the Hannover Fair. It is used as a synonym for Cyber-Physical Production Systems (CPPS), i.e., Cyber-Physical Systems applied in the domain of manufacturing/production (Vogel-Heuser and Hess 2016). The associated technologies may revolutionize the manufacturing industry by building "smart factories" that can make intelligent decisions and improve the performance of the manufacturing systems and products. From this aspect, there are several other similar terms that share the same technologies of Industry 4.0, such as smart manufacturing and intelligent manufacturing. Wang et al. (2021) conducted a comparative literature review between smart manufacturing and intelligent manufacturing in terms of their focus, components, concepts, etc.

"Industry 5.0" was introduced in 2021 by the European Commission. It "complements the existing Industry 4.0 approach by specifically putting research and innovation at the service of the transition to a sustainable, human-centric and resilient (European) industry" (Breque et al. 2021). It aims beyond efficiency and productivity as the sole goals and reinforces the role and the contribution of industry to society (including focusing on the wellbeing of the workers, respecting the production limits of the planet, etc.). A similar concept "Society 5.0" was initiated by the Japanese government in 2016 to balance economic advancement with the resolution of social problems (e.g., aged tendency, low birth rates, and lack of competitiveness). The goal of Society 5.0 is a human-centric super-smart society to ensure all citizens can access high-quality lives full of comfort and vitality by providing necessary goods and services for individual people at the necessary level when needed through merging between cyberspace and physical space with 5G, big data, artificial intelligence, etc. (Huang et al. 2022a).

Therefore, Industry 4.0 and Industry 5.0 are not two separate eras. Instead, they coexist and complement each other. While Industry 4.0 is technology-driven with a focus on using digitalization and AI-driven technologies to improve production efficiency and flexibility, Industry 5.0 is value-driven and emphasizes the societal needs and personalized value creation (Xu et al. 2021). They emphasize different aspects of the "Techno-Social System", with technology as the enabling tools and social needs as the goals (Leng et al. 2022). In this chapter, we call it "the era of Industry 4.0 and 5.0".

Regardless the term used, the current era differs from the time when RMS was invented. From the supply chain perspective, because of its development and evolution in the past three decades, the global supply network has been becoming increasingly intercoupled and complicated. From the customer demand perspective, today's customers have an increased desired for products and services that meet their individualized needs and reflect their values, which has greatly increased the number of product variants needed. All these factors make the prediction of the market demand (both product volume and functionality) significantly more complicated, as well as the operations in the manufacturing systems that can cost-effectively produce a large number of product variants. In addition, new socio-economic systems, such as sharing economy and circular economy, have changed the customer habits, and customers' role in the product design and development process. New business models and manufacturing modes, such as cloud manufacturing (i.e., manufacturing as a service) (Liu and Xu 2017), are emerging.

4.2 RMS in the Era of Industry 4.0 and 5.0

4.2.1 RMS and Industry 4.0

Boston Consulting Group identified nine key enabling technologies of Industry 4.0 (Rüßmann et al. 2015), including big data and analytics, autonomous robots, simulation, horizontal and vertical system integration, Industrial Internet of Things, cybersecurity, cloud, additive manufacturing, and augmented reality. In Bortolini et al. (2018), the "autonomous robots" technology is further complemented by a broader category of technologies called "advanced manufacturing solutions", which refer to the set of autonomous, cooperating industrial robots and refer to the class of modular manufacturing systems characterized by integrated sensors and standardized interfaces. RMS technologies are considered a category of such "advanced manufacturing technologies" and thus belong to Industry 4.0 technologies. The RMS concepts, principles, and technologies have increasingly been implemented in today's smart factories to realize their manufacturing goals and objectives.

For example, as mentioned earlier, one goal for today's manufacturing systems is to produce a large number of "market-of-one" products at a very high efficiency, enabling a "mass-individualization" manufacturing paradigm (Gu and Koren 2022). Some of the RMS principles, such as modular design of product, and building a manufacturing system that can simultaneously and cost-effectively manufacture various individualized products, can be developed based on the RMS technologies and principles. Another example is to deal with the sudden change on the supply chain. During Covid-19, many companies around the world reconfigured their existing manufacturing resources to produce healthcare products (e.g., ventilators, masks, and medical scrubs) due to the surge in the demand of these products (Napoleone and Prataviera 2020). Also, the modular structure of reconfigurable machines can be extended to design of reconfigurable industrial robots (Valente 2016) that are increasingly used in today's manufacturing systems. These examples demonstrate the value of RMSs in meeting the manufacturing goals in the current era.

Furthermore, Industry 4.0 technologies improve the performance of RMS, or the *reconfigurability*. These technologies can enhance the reconfiguration of various elements (e.g., machine, robot, material handling system) in the manufacturing systems by reducing the reconfiguration effort and time, as well as improving the reconfigurability indicators such as cost-effectiveness and diagnosability. For example, Industrial Internet of Things enables the real-time collection of the machine status data, enabling the monitoring of the machine health condition. In addition, data analytics and machine learning algorithms can be developed to discover new insights from the data and make smart decision on the production and maintenance in the system, thus improving the system diagnosability.

Napoleone et al. (2023b) provided a literature review on how technologies underlying CPSs support the reconfigurability capability in manufacturing. These CPS-related technologies are classified into seven classes: (1) sensor, measurement and data acquisition technologies; (2) communication and connectivity technologies, open and standard interfaces; (3) edge, fog and cloud computing; (4) simulation, artificial intelligence and machine learning; (5) advanced monitoring and Digital Twin; (6) ubiquitous computing, assistance systems, augmented and virtual reality, and human–machine interfaces; (7) decentralized control architecture. Among these techniques, one key technique that can facilitate the design and operation of RMS is Digital Twin (DT). Touckia (2023) reviewed the role of DT in the RMS evaluation, and proposed using it for future research on the integration of DT in the evaluation of RMS. The benefits of DT are multiple such as evaluation of current and future capabilities of an RMS during its life cycle, early discovery of system performance deficiencies and production optimization. Zhang et al. (2019) proposed a reconfigurable DT manufacturing system and reconfigurable strategies for different levels of the manufacturing system, from equipment level to service level. In addition to RMS, DT can also be used to support the design of the reconfigurable machine tools (Huang et al. 2022b).

Note that, the concept of "reconfigurability" is also evolving over the time. In addition to the six core RMS characteristics described in Sect. 3.1, Morgan et al. (2021) included the following two new characteristics: "*Adaptability—ability to be responsive to changes in production volume and product characteristics.*" and "*Mobility—ability to move products through the system* (*transport mechanisms and resources*)." "Adaptability" is similar as "responsiveness" defined in Koren (2013). "Mobility" is regarding the material handling devices in the manufacturing systems. As pointed in Gu and Koren (2022), adding flexible routing to an existing RMS may improve such mobility and the flexibility/reconfigurability of the system.

Furthermore, the integration of RMS technologies and other Industry 4.0 technologies can support the design and operations of RMS, and to improve the system reconfigurability and other performance measures (e.g., efficiency). For example, one of the most distinctive and signature keywords in Fig. 5 is *RIS* (*Reconfigurable Inspection System*), which integrates reconfigurability with intelligent sensing capabilities. By incorporating advanced sensors to capture both visual and numerical data, reconfigurable inspection machines enhance process monitoring and quality control. This fusion of reconfigurability allows for the rapid adaptation of inspection parameters, sensor configurations, and analytical methods to accommodate variations in product design, material properties, and production conditions. This transition from early reconfigurable manufacturing to smart reconfigurable manufacturing reflects a fundamental shift from the focus on modularity and control architectures to a digitally connected, AI-driven paradigm that enhances adaptability and intelligence in a more dynamic reconfigurable manufacturign environment.

4.2.2 RMS and Industry 5.0

Industry 5.0 is value-driven and has three main objectives: sustainability, resilience, and human-centricity. We provide a brief discussion below on the relationship between RMS and each of the three objectives.

RMS and Sustainability. RMS is considered as a sustainable "living" factory because its lifecycle can be extended to meet the production requirements for multiple generations of products (Koren et al. 2018). Methods and practices around "6R"-based principles (i.e., reduce, reuse, recycle, recover, redesign, and remanufacture) can be developed to improve the sustainability of the products, processes, and systems (Jawahir and Bradley 2016). Many of the RMS characteristics and principles align well with these 6R-principles and they can be well integrated with each other. RMS can improve and promote sustainability by extending the life cycle of the system itself and that of the products manufactured in the system. From the system life cycle perspective, RMS is more sustainable as it has longer life cycles. Unlike traditional DMLs where the manufacturing resources (e.g., dedicated machines) will be obsolete when the product requirement changes, the convertibility of RMS enables the system to be used for new generation of products or new product variants. Even when the system as a whole is no longer needed, modular design enables the components and modules of the RMS to be easily reused or recycled. Moreover, RMS can be used to extend the life cycle of other products through manufacturing operations and thus improving sustainability. For example, RMS can be used for cost-effective remanufacturing, where the quality of the incoming parts is highly uncertain and heterogenous, and thus different operations are required.

In addition, the "sustainability" performance can be further improved in the RMS through the optimal design and operations. Orellano et al. (2022) and Skärin et al. (2023) reviewed the literature on using sustainability as a metric (which can be evaluated from environmental, economic, and social aspects) to optimize the RMS design and operations.

RMS and Resilience. Resilience of a manufacturing system means its ability to quickly recover from potential disruptions and adaptively mitigate the negative impacts caused by disruptions. Such disruptions may occur at the machine or system level (e.g., machine failures) and the supply chain level (e.g., material shortage). Reconfiguration is considered as an important strategy to deal with these disruptions and improve the resilience. For example, Gu et al. (2015) discussed the design strategies for a resilient manufacturing system through adopting RMS configuration in the design stage and through reconfiguration in the operation stage (after disruption occurs). In addition, the diagnosability of the RMS may help to accurately predict certain disruptions (e.g., machine failures) in advance and prevent their occurrence through predictive maintenance.

In addition to the manufacturing system level, the reconfigurable manufacturing principles can also be implemented to a supply chain level to enhance the resilience of the supply chain network. Reconfigurable Supply Chain (RSC) is defined by Dolgui et al. (2020) as "a network designed in a cost-efficient, responsive, sustainable and resilient manner that is increasingly data-driven and dynamically adaptable and capable for rapid structural changes in physical and cyber spaces, by rearrangement and reallocation or change of its components in order to quickly adjust supply and production capacities and functionality in response to sudden changes." This definition clearly shows that resilience and reconfiguration are jointly addressed in the RSC design. In addition, Al Naimi et al. (2022) did a systematic mapping review on literatures on supply chain resilience and supply chain reconfiguration.

RMS and Human-centricity. Human-centricity means taking the core human needs and interests (e.g., safety, well-being, personal development, work-life balance of the workers and operators) as priorities in the manufacturing plant. To address such needs, the manufacturing systems should be adaptive to the workers' individual capabilities and needs. This can be achieved by reconfigurable technologies through reconfigurability and adaptability. For example, Li et al. (2023) developed a framework that utilizes RMS to facilitate the up skilling of workers.

The highly complex and uncertain manufacturing environment of modern manufacturing systems require to combine the high accuracy of the automated machines and robots as well as human intelligence and experience. Therefore, techniques such as human–machine collaboration and human–robot collaboration are increasingly being developed and implemented. Incorporation of these human-centricity-focused techniques into RMSs can also improve the system performance. For example, Napoleone et al. (2023a) provided a state-of-the-art review on "human-centric reconfigurable manufacturing systems", with a focus on reconfigurability enablers for systems, machines, robots, materials handling systems, and operators that can reduce reconfiguration effort in automated, mixed, and manual systems.

From the above discussion, RMS technologies are key enabling technologies to meet not only the traditional RMS goals, but also the new goals of the current era of Industry 4.0 and 5.0, such as increased product variety, smartness, sustainability, resilience, and human-centricity. These new goals can also enrich RMS technologies by providing new aspects for the design and operations of RMS. On the other hand, reconfigurability can be further improved by these new technologies and principles developed in the current era. Figure 6 shows the relation between RMS and Industry 4.0/5.0 regarding their goals and technologies.

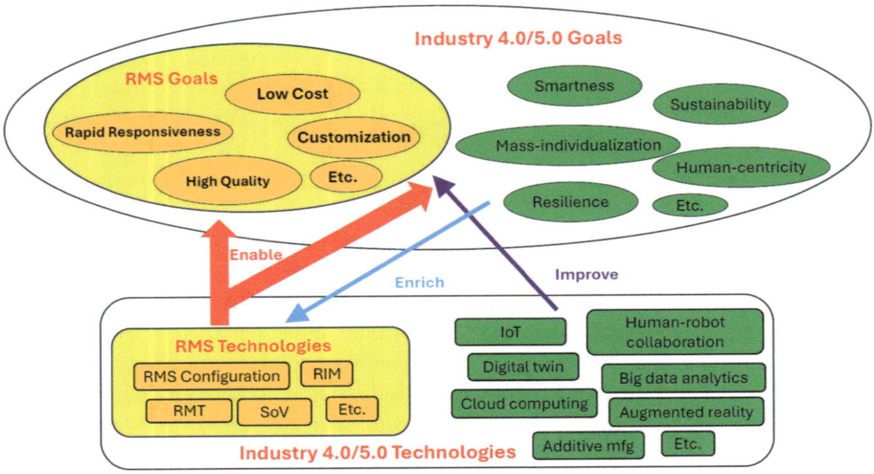

Fig. 6 Relationship between RMS goals and technologies and Industry 4.0/5.0 goals and technologies

5 Conclusions

Since the concept of RMS was proposed three decades ago, the RMS paradigm has received growing attention and interest from both academia and industry. This chapter reviews the roadmap of the RMS development, including its emergence and some of the early pioneered works at the ERC/RMS that established the scientific basis of RMS, such as RMS core characteristics, design principles, inventions, patents, and technologies. We also surveyed the RMS research topics and trend in the literature, which demonstrates a significantly increased number of publications in the current era of Industry 4.0 and 5.0.

We discuss the role of RMS in the current Industry 4.0 and 5.0 era. On the one hand, RMS principles and technologies are considered as enabling technologies for Industry 4.0 and 5.0, and to realize the goals of today's manufacturing enterprises. On the other hand, the new technologies developed in the current era, such as Internet-of-Things and Digital Twin, can improve the reconfigurability of the machines and the production systems. The integration of RMS technologies and the other technologies of Industry 4.0 and 5.0 has great potentials to revolutionize the smart factories of today and the future.

References

Al Naimi M, Faisal MN, Sobh R, Bin Sabir L (2022) A systematic mapping review exploring 10 years of research on supply chain resilience and reconfiguration. Int J Log Res Appl 25(8):1191–1218

Andersen A, Brunoe TD, Nielsen K, Rösiö C (2017) Towards a generic design method for reconfigurable manufacturing systems: analysis and synthesis of current design methods and evaluation of supportive tools. J Manuf Syst 42:179–195

Arista R, Mas F, Morales-Palma D, Vallellano C (2022) Industrial Resources in the design of reconfigurable manufacturing systems for aerospace: a systematic literature review. Comput Ind 142:103719

Bi ZM, Lang SY, Shen W, Wang L (2008a) Reconfigurable manufacturing systems: the state of the art. Int J Prod Res 46(4):967–992

Bi ZM, Lang SY, Verner M, Orban P (2008b) Development of reconfigurable machines. Int J Adv Manuf Technol 39(11):1227–1251

Bortolini M, Galizia FG, Mora C (2018) Reconfigurable manufacturing systems: literature review and research trend. J Manuf Syst 49:93–106

Breque M, De Nul L, Petridis A (2021) Industry 5.0: towards a sustainable, human-centric and resilient European industry, vol 46. European Commission, Directorate-General for Research and Innovation, Luxembourg, LU

Dolgui A, Ivanov D, Sokolov B (2020) Reconfigurable supply chain: the X-network. Int J Prod Res 58(13):4138–4163

Gadalla M, Xue D (2017) Recent advances in research on reconfigurable machine tools: a literature review. Int J Prod Res 55(5):1440–1454

Gu X, Koren Y (2022) Mass-individualisation–the twenty first century manufacturing paradigm. Int J Prod Res 60(24):7572–7587

Gu X, Jin X, Ni J, Koren Y (2015) Manufacturing system design for resilience. Procedia CIRP 36:135–140

Huang S, Wang B, Li X, Zheng P, Mourtzis D, Wang L (2022a) Industry 5.0 and Society 5.0—comparison, complementation and co-evolution. J Manuf Syst 64:424–428

Huang S, Wang G, Yan Y (2022b) Building blocks for digital twin of reconfigurable machine tools from design perspective. Int J Prod Res 60(3):942–956

Jawahir IS, Bradley R (2016) Technological elements of circular economy and the principles of 6R-based closed-loop material flow in sustainable manufacturing. Procedia CIRP 40:103–108

Koren Y (2010) The global manufacturing revolution: product-process-business integration and reconfigurable systems. Wiley & Sons

Koren Y (2013) The rapid responsiveness of RMS. Int J Prod Res 51(23–24):6817–6827

Koren Y (2022) The invention of reconfigurable manufacturing systems–a survey of RMS early patents. Manuf Lett 33:51–53

Koren Y, Shpitalni M (2010) Design of reconfigurable manufacturing systems. J Manuf Syst 29(4):130–141

Koren Y, Heisel U, Jovane F, Moriwaki T, Pritschow G, Ulsoy G, Van Brussel H (1999) Reconfigurable manufacturing systems. CIRP Ann 48(2):527–540

Koren Y, Gu X, Badurdeen F, Jawahir IS (2018) Sustainable living factories for next generation manufacturing. Procedia Manuf 21:26–36

Leng J, Sha W, Wang B, Zheng P, Zhuang C, Liu Q, Wuest T, Mourtzis D, Wang L (2022) Industry 5.0: prospect and retrospect. J Manuf Syst 65:279–295

Li X, Athinarayanan R, Wang B, Yuan W, Zhou Q, Jun M, Bravo J, Gao RX, Wang L, Koren Y (2024) Smart reconfigurable manufacturing: literature analysis. Procedia CIRP 121:43–48

Li X, Nassehi A, Wang B, Hu SJ, Epureanu BI (2023) Human-centric manufacturing for human-system coevolution in Industry 5.0. CIRP Ann 72(1):393–396

Liu Y, Xu X (2017) Industry 4.0 and cloud manufacturing: a comparative analysis. J Manuf Sci Eng 139(3):034701

Maler-Speredelozzi V, Koren Y, Hu SJ (2003) Convertibility measures for manufacturing systems. CIRP Ann 52(1):367–370
Morgan J, Halton M, Qiao Y, Breslin JG (2021) Industry 4.0 smart reconfigurable manufacturing machines. J Manuf Syst 59:481–506
Napoleone A, Andersen A, Brunoe TD, Nielsen K (2023a) Towards human-centric reconfigurable manufacturing systems: literature review of reconfigurability enablers for reduced reconfiguration effort and classification frameworks. J Manuf Syst 67:23–34
Napoleone A, Negri E, Macchi M, Pozzetti A (2023b) How the technologies underlying cyber-physical systems support the reconfigurability capability in manufacturing: a literature review. Int J Prod Res 61(9):3122–3144
Napoleone A, Prataviera LB (2020) Reconfigurable manufacturing: lesson learnt from the COVID-19 outbreak. In: Paper presented at the IFIP international conference on advances in production management systems, pp 457–465
National Science Foundation (1996) Award abstract # 9529125 engineering research center for reconfigurable machining systems. https://www.nsf.gov/awardsearch/showAward?AWD_ID=9529125. Last Accessed 20 Mar 2025
Orellano M, Sabioni RC, Pacheco A (2022) Reconfigurable manufacturing systems from a sustainability perspective: a systematic literature review. In: Paper presented at the IFIP international conference on advances in production management systems, pp 152–159
Renzi C, Leali F, Cavazzuti M, Andrisano AO (2014) A review on artificial intelligence applications to the optimal design of dedicated and reconfigurable manufacturing systems. Int J Adv Manuf Technol 72:403–418
Rüßmann M, Lorenz M, Gerbert P, Waldner M, Justus J, Engel P, Harnisch M (2015) Industry 4.0: the future of productivity and growth in manufacturing industries. Boston Consult Group 9(1):54–89
Shi J (2006) Stream of variation modeling and analysis for multistage manufacturing processes. CRC Press
Skärin F, Rösiö C, Andersen A (2023) Sustainability and circularity in reconfigurable manufacturing-literature review and future research directions. Int J Manuf Res 18(4):366–391
Touckia JK (2023) Integrating the digital twin concept into the evaluation of reconfigurable manufacturing systems (RMS): literature review and research trend. Int J Adv Manuf Technol 126(3):875–889
Valente A (2016) Reconfigurable industrial robots: a stochastic programming approach for designing and assembling robotic arms. Robot Comput-Integr Manuf 41:115–126
Vogel-Heuser B, Hess D (2016) Guest editorial Industry 4.0–prerequisites and visions. IEEE Trans Autom Sci Eng 13(2):411–413
Wang W, Koren Y (2012) Scalability planning for reconfigurable manufacturing systems. J Manuf Syst 31(2):83–91
Wang B, Tao F, Fang X, Liu C, Liu Y, Freiheit T (2021) Smart manufacturing and intelligent manufacturing: a comparative review. Engineering 7(6):738–757
Xu X, Lu Y, Vogel-Heuser B, Wang L (2021) Industry 4.0 and Industry 5.0—inception, conception and perception. J Manuf Syst 61:530–535
Yelles-Chaouche AR, Gurevsky E, Brahimi N, Dolgui A (2021) Reconfigurable manufacturing systems from an optimisation perspective: a focused review of literature. Int J Prod Res 59(21):6400–6418
Zhang C, Xu W, Liu J, Liu Z, Zhou Z, Pham DT (2019) A reconfigurable modeling approach for digital twin-based manufacturing system. Procedia CIRP 83:118–125

Intelligent Clustering Method of Part Family Formation for D-RMS

Sihan Huang, Ming Huang, Guoxin Wang, and Yan Yan

Abstract Delayed reconfigurable manufacturing systems (D-RMS), a specialized subset of reconfigurable manufacturing systems (RMS), have been introduced to address the convertibility challenges inherent in traditional RMS. An exclusive part family formation method of D-RMS based on machine learning is proposed in this chapter. Firstly, a similarity coefficient that considers the characteristics of delayed reconfiguration is constructed. The positions of common operations within their respective sequences are analyzed, with earlier common operations increasing the likelihood of parts being grouped into the same family. To further refine this analysis, the concept of the longest relative position common operation subsequence (LPCS) is introduced, which accounts for the relative positions of common operations. Additionally, the position differences and discontinuities of LPCSs within the corresponding operation sequences are examined. Secondly, K-medoids is adopted to group parts into families based on the similarity among parts. Finally, a case study is conducted to demonstrate the effect of the proposed method.

1 Introduction

The concept of a delayed reconfigurable manufacturing system (D-RMS) was introduced to smoothly address the challenge of convertibility by dividing the manufacturing system into two subsystems, allowing partial production activities to continue during reconfiguration (Huang et al. 2019). As a subclass of reconfigurable manufacturing systems (RMS) (Huang et al. 2018a), D-RMS inherits all RMS characteristics. Additionally, D-RMS aims to defer reconfiguration activities to the latest possible stages. Naturally, the implementation of D-RMS should begin with part-family formation (Koren et al. 1999; Gauss et al. 2021). Although D-RMS can

S. Huang (✉) · M. Huang · G. Wang · Y. Yan
School of Mechanical Engineering, Beijing Institute of Technology, Beijing, China
e-mail: hsh@bit.edu.cn

S. Huang et al. (eds.), *Design and Operation of Smart Reconfigurable Manufacturing Systems in Industry 4.0/5.0*, Springer Series in Advanced Manufacturing,
https://doi.org/10.1007/978-3-032-00284-6_2

ideally optimize part-family formation for RMS, neglecting its delayed reconfiguration characteristics during this process may lead to increased reconfiguration costs in practice.

In advancing generic part-family formation, Choobineh (1988) integrated operation sequences with similarity coefficients, while Tam (1990) introduced a clustering algorithm to refine their post-processing. Selim et al. (2003) further enhanced the clustering approach used in part-family classification. The effects of compatibility index and longest common subsequence (LCS) on part-family formation were explored by Askin and Zhou (1998). Mukattash et al. (2018) emphasized the efficacy of part-family formation. In the scope of RMS, Abdi and Labib (2004) analyzed the reconfiguration link during product selection. Ashraf and Hasan (2015) combined RMS attributes—modularity, commonality, compatibility, and reusability—with the analytic hierarchy process (AHP) to investigate part-family structuring. Galan et al. (2007) computed a similarity matrix for product requirements, applying the AHP methodology to generate a consolidated similarity matrix, which was then processed using the Average Linkage Clustering (ALC) algorithm to derive product families. Abdi (2012) incorporated production volume, cost, and other factors into part family computation. Goyal et al. (2013) developed an operation sequence-based BMIM (bypassing moves and idle machines) similarity coefficient using LCS, optimizing part classification through average linkage hierarchical clustering. Gaurav et al. (2024) introduced similarity coefficients derived from LCS and classified part families through average linkage hierarchical clustering in a case study. Hasan et al. (2018) addressed the complexity level during part family formation of RMS, integrating them with operation sequences using binary-rooted trees. Huang and Yan (2019) studied the influences of process time and capacity demand on RMS part family formation. The idle machines and bypass moves were analyzed by Goyal et al. (2013). Ameer and Dahane (2022) defined a new reconfiguration effort index that considers all reconstruction efforts at the process level, including changes in all configurations, machines, and fixtures. Shivdas and Sapkal (2023) presented a composite similarity metric (CSM) to determine the similarity coefficient for RMS part family formation. Wang et al. (2016) discussed the importance of LCS during part family formation of RMS. Huang et al. (2018b) were aware of the importance of the reconfiguration efforts during part family formation. Several studies highlighted clustering algorithms in part-family formation. The average linkage clustering (ALC) algorithm is the most common method (Rakesh et al. 2010; Goyal et al. 2013; Wang et al. 2016; Huang et al. 2018b). While Gupta et al. (2014) optimized results using K-means clustering. Khanna and Kumar (2017) investigated the effectiveness of bond energy algorithm when clustering parts into families in RMS. Kashkoush and ElMaraghy (2014) adopted the consensus tree-based clustering algorithm in part family formation of reconfigurable assembly systems. Table 1 summarizes the key studies, including considered factors and clustering algorithms.

Table 1 Related works summary

References	Consider factor	Clustering algorithm
Abdi and Labib (2004)	Reconfiguration link	AHP
Ashraf and Hasan (2015)	Modularity, commonality, Compatibility, reusability, and product demand	AHP
Galan et al. (2007)	Modularity, commonality, Compatibility, reusability	AHP
Abdi (2012)	Production volume, cost	Analytical network process
Goyal et al. (2013)	BMIM, LCS	Average linkage hierarchical clustering
Gaurav et al. (2024)	LCS	Average linkage hierarchical clustering
Hasan et al. (2018)	Complexity level	Hierarchical clustering
Huang and Yan (2019)	Process time and capacity demand	Netting clustering
Goyal et al. (2013)	Idle machine and bypass move	ALC
Wang et al. (2016)	LCS	ALC
Huang et al. (2018b)	Reconfiguration efforts	ALC
Rakesh et al. (2010)	–	ALC
Gupta et al. (2014)	–	K-means
Khanna and Kumar (2017)	–	Bond energy
Kashkoush and ElMaraghy (2014)	Product demand, commonality	Consensus tree

Above all, although numerous studies have explored part-family formation in RMS, few have examined the impact of postponing reconfiguration activities on this process. Most of the existing methods focus on the length of the common operation sequence (e.g., LCS). The influence of the common operations on the reconfiguration activities has rarely been rarely mentioned. Consequently, these methods struggle to construct part families that align with D-RMS characteristics. To address this gap, a dedicated part-family formation method for D-RMS, incorporating delayed reconfiguration factors, must be thoroughly investigated.

2 Problem Statement

2.1 D-RMS

D-RMS, a subset of RMS, doesn't supply every single function that might be needed. Instead, it offers a specific set of additional capabilities, ensuring that the required functionality is available exactly when it's called for. It is composed of two subsystems and a warehouse between them. Subsystem 1 fulfills all the functional requirements for the entire part family. In contrast, Subsystem 2 is configured based on the functional needs of the current part and can be quickly reconfigured to accommodate new parts within the same part family. During the reconfiguration of the system, the warehouse situated between Subsystem 1 and Subsystem 2 enables the continuation of production activities for semi-finished parts. By doing so, it effectively mitigates production losses that would otherwise occur. Moreover, the redundant functions within Subsystem 1 are beneficial as they lead to a reduction in reconfiguration activities. This reduction not only cuts down on the system ramp-up time but also enhances the responsiveness of D-RMS. When considering part family, D-RMS has the ability to handle all kinds of part families of RMS. To optimize the benefits that D-RMS can offer, it is highly recommended to construct exclusive part family that incorporates the characteristics of delayed reconfiguration. Clearly, the delayed product differentiation is a special case of part family for D-RMS. Figure 1a and b respectively display an example of delayed product differentiation and its D-RMS configuration.

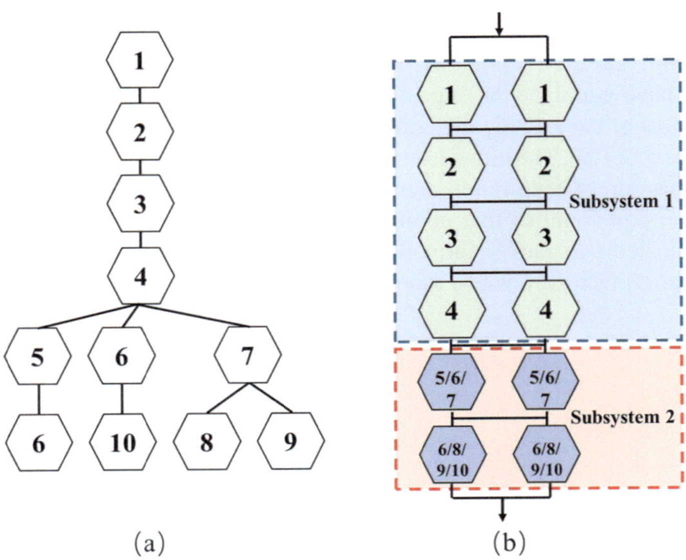

Fig. 1 Part family with delayed differentiation characteristics and its configuration

2.2 Problem Analysis

Even though part family with delayed differentiation features is an optimal solution for D-RMS, it is given from the part design viewpoint. From the standpoint of part family formation, it is essentially a process of classifying existing parts into part families, however, these parts may not inherently possess the delayed differentiation characteristics. Thus, to construct the exclusive part family with delayed reconfiguration characteristics for D-RMS, several problems should be solved.

D-RMS provides all the functions required by the part family in Subsystem 1. To reduce the function redundancy within Subsystem 1, it is essential to identify parts that share former common operations. The greater the number of such former common operations between parts, the higher the probability that they should be grouped into the same part family. For instance, consider part *m*, part *n* and part *p* with operation sequences $\{1, 2, 3, 4, 5\}$, $\{1, 2, 6, 7, 8\}$, and $\{2, 6, 7, 4, 5\}$ respectively, as depicted in Fig. 2. The common operations of part *m* and part *n* are operation $\{1\}$ and operation $\{2\}$, which are located in the former positions of their respective operation sequences. However, the most common operations of part *m* and part *p* (operation $\{2\}$, operation $\{4\}$, and operation $\{5\}$) are situated in the latter positions of their operation sequences. Although part *m* and part *p* have more common operations in total, it is more suitable for D-RMS to group part *m* and part *n* together. This is because the functional redundancy of the part family {part *m*, part *n*} is lower than that of the part family {part *m*, part *p*}, as clearly shown in Fig. 3a and b.

Common operations are sensitive to their positions within the operation sequences. On the one hand, the relative positions among these common operations will have an

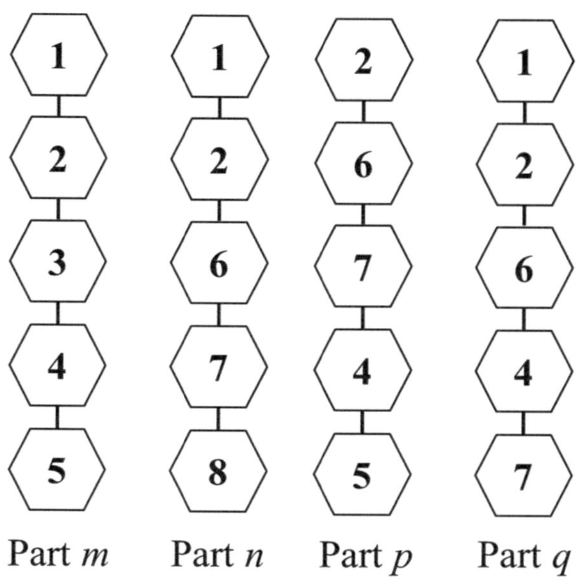

Fig. 2 Part *m*, part *n* and part *p* and the corresponding operation sequences

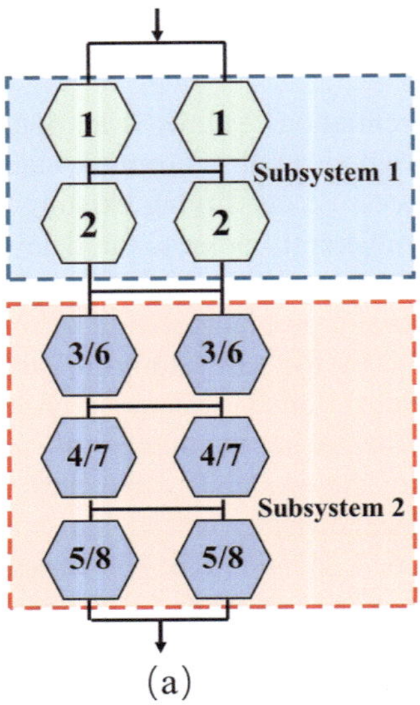

Fig. 3 **a** System configuration of part *m* and part *n*; **b** System configuration of part *m* and part *p*

impact on function redundancy. Take part *m* and part *p* as an example. In part *m*, there is one non-common operation between common operation {2} and common operation {4}, while in part *p*, there are two non-common operations between the same pair of common operations. Due to these differences in relative positions, the function of operation {2} appears two times, as illustrated in Fig. 3b. On the other hand, the different positions of common operations within their own operation sequence also affect function redundancy. For part *n* and part *p*, their common operations are operation {2}, operation {6}, and operation {7}. The positions of these common operations in part *n* are 2nd, 3rd and 4th respectively, whereas in part *p*, they are the 1st, 2nd and 3rd. These differences in positions result in an increased need for function redundancy, as can be seen in Fig. 4, where the function of operation {1} is not required when producing part *p*.

The discontinuity of common operations should be concerned, as it will increase function redundancy of D-RMS. For example, consider part *q* with operation sequence {1, 2, 6, 4, 7}, as shown in Fig. 2. The common operation between part *m* and part *q* are operation {1}, operation {2} and operation {4}. Although the positions of the common operation are the same, there is one non-common operation {6}/{3} between common operation {2} and common operation {4}, which increases function redundancy, as shown in Fig. 5.

Furthermore, the discontinuity of common operations should not be overlooked, as it will increase the functional redundancy of D-RMS. For example, consider part *q* with an operation sequence of {1, 2, 6, 4, 7}, as shown in Fig. 2. The common operations between part *m* and part *q* are operation {1}, operation {2}, and operation {4}. Even though the positions of these common operations are the same, there is

Fig. 4 System configuration of part n and part p

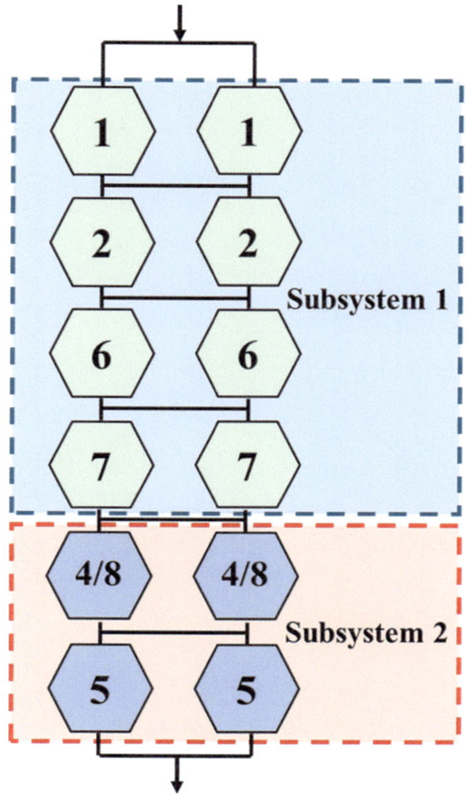

one non-common operation {6}/{3} between common operation {2} and common operation {4}. This discontinuity leads to an increase in functional redundancy, as depicted in Fig. 5.

3 Part Family Formation Method for D-RMS

In the existing literature (Wang et al. 2016), two main categories of part family formation method can be found: matrix-based methods and similarity coefficient-based methods. Among them, the similarity coefficient—based approach is chosen in this paper because of its flexibility. As a result, in the first part of this section, a similarity coefficient is put forward, which takes into account the characteristics of D-RMS. Subsequently, a clustering algorithm is introduced to group parts into families according to the calculation results of the similarity coefficient.

Fig. 5 System configuration of part m and part q

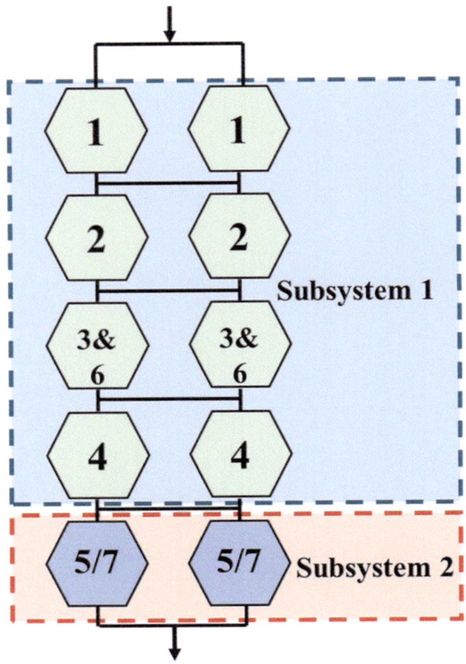

3.1 Similarity Coefficient

Typically, the degree of similarity between parts is represented by their common operations. The objective of part family formation is to group parts with more common operations into the same part family. Many studies (Askin and Zhou 1998; Goyal et al. 2013; Wang et al. 2016; Huang and Yan 2019) have utilized LCS derived from common operations to indicate the similarity degree between parts. However, the LCS fails to consider the relative positions of common operations, which should be concerned in the part family formation of D-RMS. Consequently, a subsequence of the LCS that takes into account the relative positions of common operations, named LPCS, is proposed. The LPCS is designed to address the issue of relative position problem. For example, for part m and part p, the $LCS_{mp} = \{2, 4, 5\}$. Nevertheless, since the relative positions of operation $\{2\}$, $\{4\}$, and $\{5\}$ vary, the LPCS of part m and part p is $LPCS_{mp} = \{4, 5\}$, as illustrated in Fig. 6. To identify parts with more former common operations, the reciprocal of the operation position is used as the characteristic value of LPCS. For instance, the characteristic value sequence of the LPCS of part m and part p is $\left\{\frac{1}{4}, \frac{1}{5}\right\}$.

Considering that the different positions of LPCSs will increase the function redundancy of D-RMS, the position difference degree of LPCSs should be recognized during similarity calculation. For example, the LPCS elements of part n and part p ($\{2, 6, 7\}$) are in different positions in their respective operation sequences, as shown

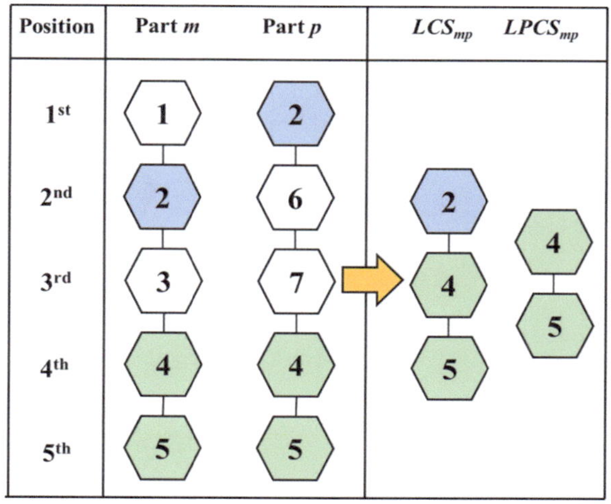

Fig. 6 LCS and LPCS of part m and part p

in Fig. 7. Besides, the discontinuity of LPCSs should be investigated to remove the part with too many non-common operations, for example the non-common operation {3}/{6} among the LPCS of part m and part q shown in Fig. 8. Above all, similarity coefficient that incorporates the lengths of the operation sequences of parts, the characteristic value sequence of LPCS, the position differences and the discontinuity of LPCSs is proposed as shown in Eq. (1).

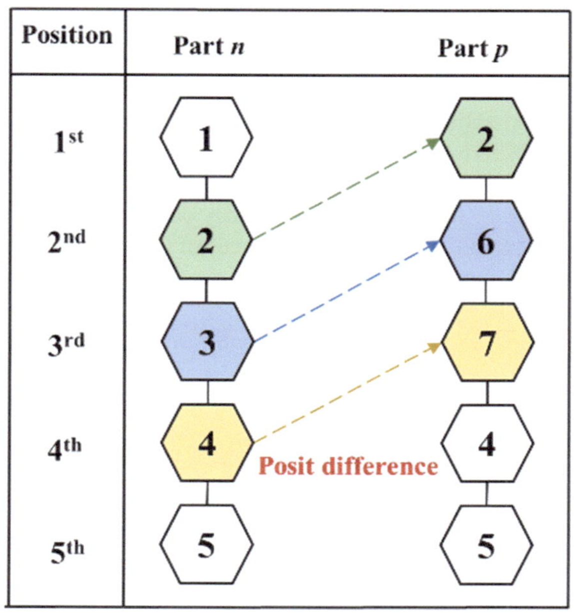

Fig. 7 Position differences in the LPCS of part n and part p

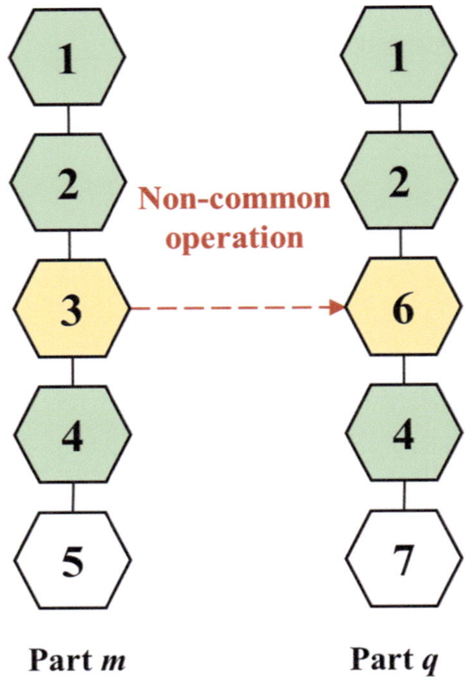

Fig. 8 Non-common operations in the LPCS of part m and part q

Considering that the different positions of LPCSs will increase the function redundancy of D-RMS, it is essential to identify the position difference degree of LPCSs during similarity calculation. For example, as shown in Fig. 7, the elements of the LPCS of part n and part p ($\{2, 6, 7\}$) are located at different positions in their respective operation sequences. Additionally, it is necessary to investigate the discontinuity of LPCSs to exclude parts with too many non-common operations. For example, there is a non-common operation $\{3\}/\{6\}$ in the LPCS of part m and part q, as depicted in Fig. 8. In summary, a similarity coefficient is proposed, which incorporates the lengths of the operation sequences of parts, the characteristic value sequence of the LPCS, the position differences, and the discontinuity of LPCSs. The formula is as follows:

$$SI_{mn} = \frac{LPCS_{mn}^{PR}}{CVS_{mn}^{PC} + \delta + \varepsilon} \tag{1}$$

where SI_{mn} is the similarity coefficient of part m and part n; $LPCS_{mn}^{PR}$ is the sum of the characteristic value sequence of $LPCS_{mn}$; CVS_{mn}^{PR} is the sum of the characteristic value sequences of part m and part n; δ is the penalty factor for the position difference; ε is the penalty factor for the discontinuity of LPCS.

CVS_{mn}^{PR} can be calculated using the following Eqs. (2)–(4).

$$CVS_{mn}^{PR} = CVS_{m}^{PR} + CVS_{n}^{PR} \tag{2}$$

$$CVS_{m}^{PR} = \sum_{i=1}^{OS_m} \frac{1}{i} \tag{3}$$

$$CVS_{n}^{PR} = \sum_{j=1}^{OS_n} \frac{1}{j} \tag{4}$$

where CVS_{m}^{PR} is the sum of the characteristic value sequences of part *m*; CVS_{n}^{PR} is the sum of the characteristic value sequences of part *n*; OS_m is the length of the operation sequence of part *m*; OS_n is the length of the operation sequence of part *n*.

$LPCS_{mn}^{PR}$ can be calculated using Eqs. (5)–(9).

$$LPCS_{mn}^{PR} = LPCS_{m}^{PR} + LPCS_{n}^{PR} \tag{5}$$

$$LPCS_{m}^{PR} = \sum_{i=1}^{OS_m} \sigma_i \frac{1}{i} \tag{6}$$

$$LPCS_{n}^{PR} = \sum_{j=1}^{OS_n} \sigma_j \frac{1}{j} \tag{7}$$

$$\sigma_m = \begin{cases} 1, & if\ \ mth\ operation\ is\ in\ LPCS_{mn} \\ 0, & others \end{cases} \tag{8}$$

$$\sigma_n = \begin{cases} 1, & if\ \ nth\ operation\ is\ in\ LPCS_{mn} \\ 0, & others \end{cases} \tag{9}$$

where $LPCS_{m}^{PR}$ is the sum of the characteristic value sequence of $LPCS_{mn}$ in part *m*; similarly, $LPCS_{n}^{PR}$ is the sum of the characteristic value sequence of $LPCS_{mn}$ in part *n*.

δ can be calculated using Eqs. (10)–(12).

$$\delta = \left| \sum_{i=1}^{OS_m} \mu_i^{min} \frac{1}{i} - \sum_{j=1}^{OS_n} \mu_j^{min} \frac{1}{j} \right| \tag{10}$$

$$\mu_i^{\min} = \begin{cases} 1, & if\ \ ith\ operation\ is\ first\ operation\ in\ LPCS_{mn} \\ 0, & others \end{cases} \tag{11}$$

$$\mu_j^{\min} = \begin{cases} 1, & if\ \ jth\ operation\ is\ first\ operation\ in\ LPCS_{mn} \\ 0, & others \end{cases} \tag{12}$$

ε can be calculated using Eqs. (13)–(19).

$$\varepsilon = \left(\sum_{i=B_m^{lower}}^{B_m^{upper}} \frac{1}{i} + \sum_{j=B_n^{lower}}^{B_n^{upper}} \frac{1}{j} \right) - LPCS_{mn}^{PR} \tag{13}$$

$$B_m^{lower} = \sum_{i=1}^{OS_m} \mu_i^{\min} i \tag{14}$$

$$B_m^{upper} = \sum_{i=1}^{OS_m} \mu_i^{\max} i \tag{15}$$

$$B_n^{lower} = \sum_{j=1}^{OS_n} \mu_j^{\min} j \tag{16}$$

$$B_n^{upper} = \sum_{j=1}^{OS_n} \mu_j^{\max} j \tag{17}$$

$$\mu_i^{\max} = \begin{cases} 1, & \text{if ith operation is first operation in } LPCS_{mn} \\ 0, & \text{others} \end{cases} \tag{18}$$

$$\mu_j^{\max} = \begin{cases} 1, & \text{if jth operation is first operation in } LPCS_{mn} \\ 0, & \text{others} \end{cases} \tag{19}$$

where B_m^{lower} and B_m^{upper} are the lower and upper position bounds of $LPCS_{mn}$ in part m separately; similarly, B_n^{lower} and B_n^{upper} are the lower and upper position bounds of $LPCS_{mn}$ in part n separately.

3.2 Clustering Algorithm for Part Family Formation

Based on the proposed similarity coefficient, a similarity matrix can be obtained. Subsequently, a clustering algorithm ought to be selected to group part into family. Hence, this section will investigate the clustering algorithm applicable to the part family formation of D-RMS.

Hierarchical clustering algorithm, which includes agglomerative and divisive algorithms, is a widely used grouping technology. For example, the Average Linkage Clustering (ALC) algorithm has been employed in many studies (Rakesh et al. 2010; Goyal et al. 2013; Wang et al. 2016; Huang et al. 2018b). However, in the era of intelligent manufacturing, the hierarchical clustering algorithm is becoming less popular and may even be ineffective in some situations. In recent years, machine learning (Xu et al. 2020; Zerrougui et al. 2022) has gained significant popularity in clustering applications. For instance, Gupta et al. (2014) attempted to optimize part family formation using K-means. Nevertheless, K-means does not necessarily select one of the input data points as the center of a cluster, which may lead to meaningless cluster centers. K-medoids (Park and Jun 2009; Ezzat et al. 2022) is a clustering algorithm similar to K-means. In contrast to K-means, K-medoids makes actual data points as centers, which makes the cluster centers more interpretable. Moreover, K-medoids can be used in conjunction with various dissimilarity measures, which is consistent with the proposed similarity coefficient. Therefore, K-medoids is adopted as the clustering algorithm for part family formation in D-RMS. The procedure of K-medoids is as follows:

Step 1. Take the similarity matrix of y parts as input.
Step 2. Randomly select x parts ($x < y$) as the medoids that is the initial clusters.
Step 3. Assign the remaining y–x parts to the closest medoid separately.
Step 4. Update the medoid of every cluster according to the following criterion:

$$\begin{cases} I = \sum_{i=1}^{y} \sum_{j=1, pa_i \in C_j}^{x} d\left(pa_i, c_j\right) \\ d\left(pa_i, c_j\right) = 1 - S_{pa_i c_j} \end{cases} \tag{20}$$

where c_j represents the jth cluster medoid, $d(pa_i, c_j)$ is the dissimilarity between part i and jth cluster medoid c_j, which can be calculated according to the proposed similarity coefficient. The part with the minimum criterion value will be set as new medoids.
Step 5. Terminal condition: Repeat Steps (3) to (4) until all medoids remain unchanged or the maximum iteration limit is reached.
Step 6. Obtain the final x clusters, that is, x part families of D-RMS.

The flowchart of the clustering process based on K-medoids is shown in → Fig. 9. It is an unsupervised learning process where no preset labels are provided before training. K-medoids aims to learn the similarity relationships among all parts. Based on these relationships, it divides all the parts into x clusters by assigning the remaining y–x parts to the closest medoids until all the medoids no longer change.

4 Case Study

In this section, a case study is present including the proposed part family formation method implementation and effectiveness verification. The details of the proposed part family formation method implementation are elaborated firstly. The effectiveness of the proposed part family formation method is discussed based on the implementation results. Then, a clustering effectiveness comparison is present to show the advantage of the adopted K-medoids.

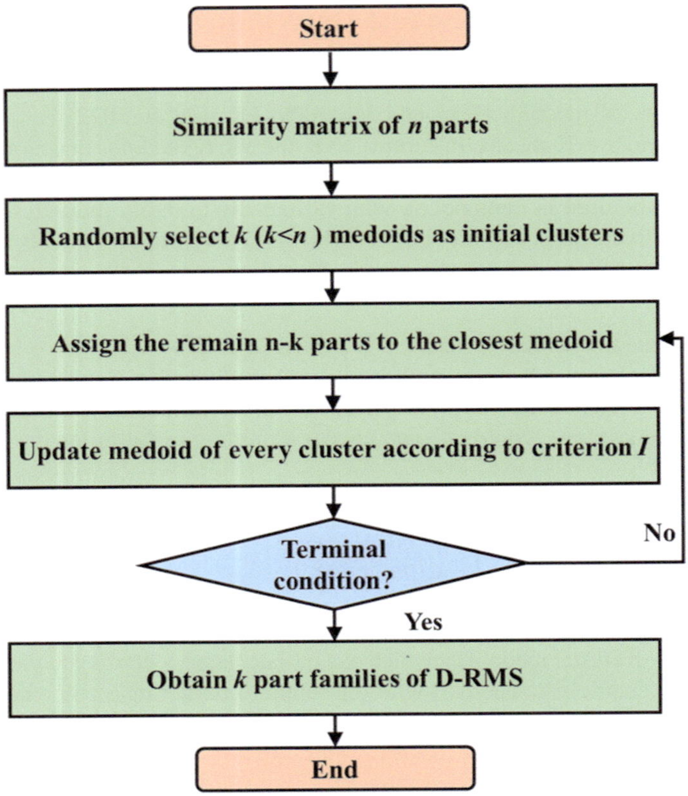

Fig. 9 The flowchart of clustering process based on K-medoids

To further validate the effectiveness of the proposed part family formation method in real-world production, this section presents a detailed analysis of an industrial case involving seven parts, as shown in Fig. 10. The operation sequences of each part involve different process steps, reflecting the diverse demands in actual production. By applying the proposed method, we can group these parts into different part families based on the similarity of their operation sequences, thus supporting the optimized configuration of the D-RMS.

The operation sequences for the seven parts in the industrial case are shown in Table 2. The lengths of these operation sequences vary from 2 to 4 steps, and the common operation positions include all scenarios: front-end common operations (e.g., Part 1 and Part 2, Part 3 and Part 6), middle common operations (e.g., Part 5 and Part 6), and back-end common operations (e.g., Part 2 and Part 3). Additionally, there are differences in the common operation positions (e.g., Part 1 and Part 5, Part 2 and Part 5), as well as discontinuities in LPCS (Local Process Similarity) (e.g., Part 13 and Part 16). Based on the input data from Table 2, the proposed method is used to group these seven parts into families. The KNIME platform serves as the

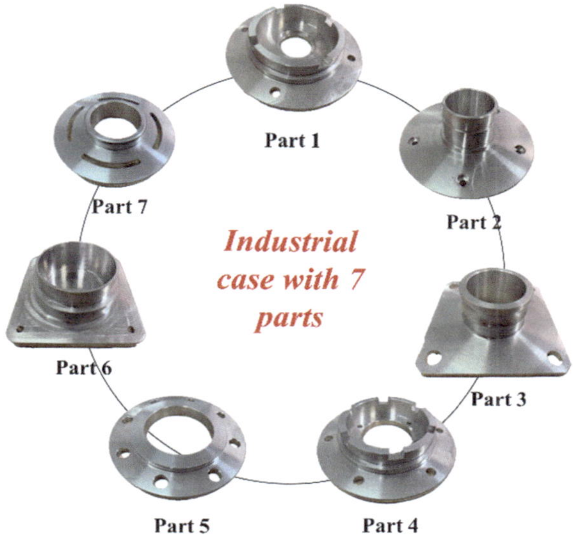

Fig. 10 Industrial case with 7 parts

experimental platform for clustering the parts. Using the operation sequences from Table 2, a similarity matrix is derived with the proposed similarity coefficient, and the calculation results are shown in Table 3.

To perform reasonable clustering of the parts, we first need to calculate the similarity between them. By using the custom similarity coefficient, we can quantify the common operations between different parts in their operation sequences. This similarity coefficient not only considers the positions of common operations but also accounts for differences in the order of operations and discontinuities.

The similarity calculation results are presented in the similarity matrix (Table 3), which shows the relationships between the various parts. This matrix allows us to identify which parts share similar process steps and which parts differ significantly from each other.

Table 2 Operation sequences of the 7 parts

Part no.	Operation sequence
1	Cylindrical turning (1), Radial turning (2), Hole drilling (3), Slot milling (4)
2	Cylindrical turning (1), Radial turning (2), Hole drilling (3), Hole boring (5)
3	Face milling (6), Hole drilling (3), Hole boring (5)
4	Cylindrical turning (1), Radial turning (2), Hole drilling (3), Slot milling (4)
5	Radial turning (2), Hole drilling (3)
6	Face milling (6), Hole drilling (3), Cylindrical turning (1)
7	Face milling (6), Slot milling (4)

Table 3 The similarity calculation results of the 7 parts

Part	1	2	3	4	5	6	7
1	–	0.880	0.204	1.000	0.571	0.291	0.196
2		–	0.347	0.880	0.571	0.291	0.000
3			–	0.204	0.300	0.818	0.600
4				–	0.571	0.291	0.196
5					–	0.300	0.000
6						–	0.600
7							–

As can be seen from Table 3, Parts 1 and 4 have identical operation sequences, resulting in a similarity coefficient of 1. The similarity coefficient between Parts 5 and 7 is 0 because they have no common operations. Excluding 1, the highest similarity coefficient of 0.880 appears between Parts 1 and 2, as well as between Parts 2 and 4. According to the operation sequences in Table 2, Parts 1 and 2 have only one different operation in their last position, and the first three operations ({1, 2, 3}) form $LPCS_{12}$. Moreover, there is no positional difference or discontinuity between $LPCS_1$ and $LPCS_2$. Similarly, Parts 2 and 4 share the first three operations ({1, 2, 3}), and the only difference is in their final operations. The second highest similarity coefficient of 0.818 occurs between Parts 3 and 6. Their first two operations ({6, 3}) form $LPCS_{36}$, with only the final operation differing between Parts 3 and 6. Overall, parts with more common operations at the beginning and smaller positional differences, without discontinuities, tend to have higher similarity coefficients. To further verify this, we observe the smallest non-zero similarity coefficient among the 20 parts, which is 0.196 between Parts 1 and 7. $LPCS_{17}$ consists of only one operation, and its position differs from other operations in the sequence. This shows that the proposed similarity coefficient correctly reflects the position of common operations in the corresponding operation sequences, with more similar operations exhibiting higher similarity coefficients. This is a significant feature of D-RMS.

The clustering algorithm used (K-medoids) is executed based on the similarity calculation results. Initial "medoids" need to be preset before clustering. In this case, $K = 2$ is set (which can be easily adjusted if needed). The original clustering results executed on the KNIME platform are shown in Fig. 11, clearly displaying the part numbers, distance matrix, distances between parts, medoids, and clustering results. For easier reading, the clustering results are reorganized in Table 4. According to Table 4, the 7 parts are grouped into 2 part families: Part family 1 = {Part 1, Part 2, Part 4, Part 5}, Part family 2 = {Part 3, Part 6, Part 7}.

By analyzing the system configuration of these two part families, we can see that the clustering results reflect the operational characteristics of each part. Parts in family 1 mainly involve early operations such as cylindrical turning and radial turning, resulting in a high degree of similarity. Therefore, they can share the initial processing resources of the production line. By grouping these parts into one family, we can

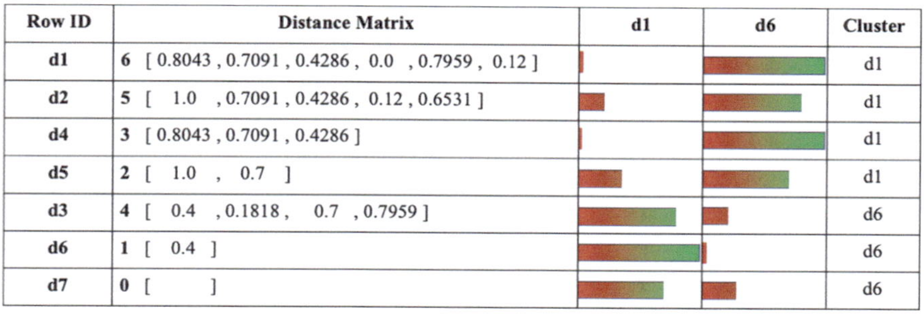

Fig. 11 Clustering results of industrial case

Table 4 Reorganized clustering results

Family no.	Medoids	Clustering results
1	Part 1	Part 1, Part 2, Part 4, Part 5
2	Part 3	Part 3, Part 6, Part 7

maximize resource utilization on the production line and reduce the complexity of scheduling and operation.

Parts in family 2, although differing in early processes, share a high degree of similarity in later operations involving face milling and slot milling. Thus, they are more suitable for sharing the same resources in the later stages of the production line. This method allows for more efficient task distribution in the production process. Clearly, these two examples embody the characteristics of D-RMS. Therefore, the proposed part family formation method can create exclusive part families for D-RMS. By dividing the parts into families, the D-RMS system can achieve flexible resource scheduling and task allocation. The characteristics of each part family determine their optimal configuration on the production line. In a multi-robot collaborative production environment, parts in family 1 can be handled by one robot, while parts in family 2 can be managed by another robot. This distribution minimizes interference between robots and improves overall production efficiency.

5 Conclusion

Although D-RMS, as a subclass of RMS, can handle all RMS part families, its postponing reconfiguration characteristics demand an exclusive part family to maximize its advantages. A similarity coefficient considering LPCS was put forward. Three major factors including former common operations, relative position of common operations, position differences and discontinuity of LPCS were analyzed. The position reciprocal was adopted as characteristic value of LPCS to highlight the former common operations. The K-medoids clustering algorithm was adopted to ensure the clustering effectiveness and efficiency of the part family formation of D-RMS. A case study validated the method's ability to meet D-RMS requirements, forming an exclusive part family.

References

Abdi MR (2012) Product family formation and selection for reconfigurability using analytical network process. Int J Prod Res 50(17):4908–4921
Abdi MR, Labib AW (2004) Grouping and selecting products: the design key of reconfigurable manufacturing systems (RMSs). Int J Prod Res 42(3):521–546
Ameer M, Dahane M (2022) Reconfiguration effort based optimization for design problem of reconfigurable manufacturing system. Procedia Comput Sci 200:1264–1273
Ashraf M, Hasan F (2015) Product family formation based on multiple product similarities for a reconfigurable manufacturing system. Int J Model Oper Manag 5(3–4):247–265
Askin RG, Zhou M (1998) Formation of independent flow-line cells based on operation requirements and machine capabilities. IIE Trans 30(4):319–329
Choobineh F (1988) A framework for the design of cellular manufacturing systems. Int J Prod Res 26(7):1161–1172
Ezzat O, Medini K, Boucher X et al (2022) A clustering approach for modularizing service-oriented systems. J Intell Manuf 33:719–734
Galan R, Racero J, Eguia I, Garcia JM (2007) A systematic approach for product families formation in reconfigurable manufacturing systems. Robot Comput-Integr Manuf 23(5):489–502
Gaurav K, Goyal KK, Batra NK et al. (2024) Execution of revised BMIM similarity coefficient for part family formation in reconfigurable manufacturing system. J Adhes Sci Technol 1–23
Gauss L, Lacerda DP, Cauchick Miguel PA (2021) Module-based product family design: systematic literature review and meta-synthesis. J Intell Manuf 32(1):265–312
Goyal KK, Jain PK, Jain M (2013) A comprehensive approach to operation sequence similarity based part family formation in the reconfigurable manufacturing system. Int J Prod Res 51(6):1762–1776
Gupta A, Jain PK, Kumar D (2014) Part family formation for reconfigurable manufacturing system using K-means algorithm. Int J Internet Manuf Serv 25 3(3):244–262
Huang S, Wang G, Shang X et al (2018a) Reconfiguration point decision method based on dynamic complexity for reconfigurable manufacturing system (RMS). J Intell Manuf 29(5):1031–1043
Hasan SM, Baqai AA, Butt SU et al (2018b) Product family formation based on complexity for assembly systems. Int J Adv Manuf Technol 95(1):569–585
Huang S, Yan Y (2019) Part family grouping method for reconfigurable manufacturing system considering process time and capacity demand. Flex Serv Manuf J 31(2):424–445
Huang S, Wang G, Yan Y et al (2018) Similarity coefficient of RMS part family grouping considering reconfiguration efforts. IEEE Access 6:71871–71883
Huang S, Wang G, Yan Y (2019) Delayed reconfigurable manufacturing system. Int J Prod Res 57(8):2372–2391
Kashkoush M, ElMaraghy H (2014) Product family formation for reconfigurable assembly systems. Procedia CIRP 17:302–307
Khanna K, Kumar R (2017) Part family and operations group formation for RMS using bond energy algorithm. Int J Eng Technol 9(2):1365–1373
Koren Y, Heisel U, Jovane F et al (1999) Reconfigurable manufacturing system. CIRP Ann-Manuf Technol 48(2):527–540
Mukattash A, Dahmani N, Al-Bashir A et al (2018) Comprehensive grouping efficacy: a new measure for evaluating block-diagonal forms in group technology. Int J Ind Eng Comput 9(1):155–172

Park HS, Jun CH (2009) A simple and fast algorithm for K-medoids clustering. Expert Syst Appl 36(2):3336–3341
Rakesh K, Jain PK, Mehta NK (2010) A framework for simultaneous recognition of part families and operation groups for driving a reconfigurable manufacturing system. Adv Prod Eng & Manag J 5(1):45–58
Selim HM, Aal RMA, Mahdi AI (2003) Formation of machine groups and part families: a modified SLC method and comparative study. Integr Manuf Syst 14(2):123–137
Shivdas R, Sapkal SU (2023) Proposed composite similarity metric method for part family formation in reconfigurable manufacturing system. Int J Adv Manuf Technol 125:2535–2548
Tam KY (1990) An operation sequence based similarity coefficient for part families formation. J Manuf Syst 9(1):55–67
Wang GX, Huang SH, Shang XW et al (2016) Formation of part family for reconfigurable manufacturing systems considering bypassing moves and idle machines. J Manuf Syst 41:120–129
Xu H, Liu Q, Casillas J et al (2020) Prediction of cell viability in dynamic optical projection stereolithography-based bioprinting using machine learning. J Intell Manuf 1–11
Zerrougui R, Adamou-Mitiche AB, Mitiche L (2022) A novel machine learning algorithm for interval systems approximation based on artificial neural network. J Intell Manuf 1–14

A NSGA-II-Based Optimization Method for Configuration Design of D-RMS

Sihan Huang, Guangyu Mo, Guoxin Wang, and Yan Yan

Abstract The core step in implementing D-RMS is configuration design, where the dividing point decision is crucial. A multi-objective optimization method for the configuration design of D-RMS is proposed by optimizing the dividing point based on NSGA-II. First, a multi-objective optimization model is formulated by considering dedicated, flexible, and reconfigurable machine tools, incorporating factors such as machine tool investment costs, reconfiguration costs, reconfiguration time, and delayed reconfiguration. The dividing point serves as the decision variable and must be an integer, making this a multi-objective integer programming problem. Second, the Non-Dominated Sorting Genetic Algorithm-II (NSGA-II) is employed to solve the proposed optimization model. Finally, a case study is conducted to validate the proposed configuration design method.

1 Introduction

In Industry 4.0 era (Lasi et al. 2014; Sigov et al. 2022), intelligent manufacturing (Zhou et al. 2018; Zhong et al. 2017; Kusiak 2018; Meng and Song 2022) is driving transformation with the support of big data (Tao et al. 2018; Fang et al. 2021), artificial intelligence (Nti et al. 2021; Yu and Gong 2024), digital twin (Lu et al. 2020; Wang et al. 2024), etc., where diversity, customization, and uncertainty have become defining characteristics of market demand that requires manufacturing systems to exhibit high adaptability and flexibility. Reconfigurable manufacturing systems (RMS) (Koren et al. 2018a; Singh et al. 2017; ElMaraghy 2005; Bortolini et al. 2018) is the most promising paradigm for handling dynamic market demands based on its cost-effective reconfiguration. By integrating with advanced information technologies (Tao and Qi 2017; Zhou et al. 2020, Zhou and Tao 2023), RMS can enhance production efficiency and intelligence. To further improve the adaptability of RMS, Huang et al. (2019) proposed Delayed Reconfigurable Manufacturing

S. Huang (✉) · G. Mo · G. Wang · Y. Yan
School of Mechanical Engineering, Beijing Institute of Technology, Beijing, China
e-mail: hsh@bit.edu.cn

S. Huang et al. (eds.), *Design and Operation of Smart Reconfigurable Manufacturing Systems in Industry 4.0/5.0*, Springer Series in Advanced Manufacturing,
https://doi.org/10.1007/978-3-032-00284-6_3

System (D-RMS) by investing in additional functional capabilities to increase the convertibility.

RMS/D-RMS is constructed around a specific part family (Huang and Yan 2019) that is the basis of configuration design. Generally, the reconfiguration activities are executed to handle the demand fluctuation, which refer to scalability (Koren 2017, 2018b) and convertibility (Koren 2013; Huang et al. 2019) of manufacturing system. In mass customization era, the production volume change is the mainstream of demand fluctuation, the scalability of RMS was concerned when configuration design (Koren 2017, 2018b; Moghaddam et al. 2018, etc.). Generally, in Industry 4.0 era, the demand fluctuation shifts from production volume change to part variety change requiring convertibility of manufacturing system. The investigation of configuration design attracted many attentions of active researchers. Koren et al. (2017) analyzed the value creation of RMS scalability and proposed a design method for scalability of RMS. Moghaddam et al. (2018) addressed the configuration design problem considering a case of single-product flow line to handle demand fluctuations, which is extended to part family in their successor work (Moghaddam et al. 2020). Mansour et al. (2023) proposed an approach that integrating simultaneously the RMT layout problem with the scalable RMS. Except concerning scalability, many other factors were considered during configuration design of RMS. Shang et al. (2021) proposed a key feature-based configuration design method for RMS by considering the influence of inspection process. Huang et al. (2020) considered the part family formation and machine selection simultaneously to design configuration under delayed reconfiguration constraint. Gauss et al. (2019) attempted to solve the configuration design problem from machine level to design modular machine families based on module-based machinery. Based on integer linear programming (ILP) formulation, Rezaee and Moghaddam (2025) developed a new predictive approach to design the system configuration during all production periods based on estimated demand data. Also, some researchers focused on the configuration selection/evaluation to promote the efficiency and effectiveness of configuration design. Goyal et al. (2012) proposed a configuration selection method by measuring the machine reconfigurability and operational capability of reconfigurable machine tools. Ashraf and Hasan (2018) proposed a framework for configuration selection of a reconfigurable manufacturing flow line with the consideration of operation constraints. Pal Singh et al. (2020) tried to explore a new metric involving nine industrially relevant important factors for product flow configuration selection of RMS. Wang et al. (2017) proposed a PROTHMEE-based configuration evaluation method considering key characteristics of RMS. Similarly, Benderbal et al. (2018) proposed a configuration assessment method considering modularity. Puik et al. (2017) assessed the configuration design scheme of RMS from the perspectives of resources and lead time. Andersen et al. (2017) attempted to extract generic design method of RMS by analyzing and synthesizing of current methods. Zhang et al. (2023) proposes a novel framework that can apply to scenarios with different failure distributions to explore how configuration orders impact reliability performance. Cerqueus and Delorme (2023) propose the first combinatorial definition and a new measure to fully assess the scalability of a system at the design phase.

The D-RMS is a subclass of RMS focusing on convertibility (Huang et al. 2019), which postpones the reconfiguration activities to the latter stages of manufacturing system through additional functions investment in the former stages. The dividing point serves as a key position in the D-RMS configuration, distinguishing the non-reconfiguration area (production area) from the reconfiguration area. As a critical step in D-RMS implementation, configuration design relies on the proper determination of the dividing point to ensure its successful completion. Huang and Yan (2020) attempted to study the configuration design of D-RMS by considering the part family formation and machine selection simultaneously.

However, it will be more reasonable to determine the dividing point through optimization by considering factors associating with delayed reconfiguration, including investment cost of machine tools, reconfiguration activities, and so on. Moreover, it is effective and popular to solve configuration design problem through bi/multi-objective optimization (Bensmaine et al. 2013; Dou et al. 2020; Bortolini et al. 2021, etc.). Yelles-Chaouche et al. (2020) reviewed the configuration design studies from optimization perspectives. Battaïa et al. (2017) presents a decision support tool for designing configuration of RMS through a combinatorial optimization model. Zhang et al. (2021) proposed a required moment set based optimization approach to support the configuration design of RMS. Generally, there is more than one objective during configuration design optimization. Bensmaine et al. (2013) adopted NSGA to solve the optimization problem of RMS by selecting optimal machines selection during configuration design. Dou et al. (2016) proposed a Bi-objective optimization method to integrate configuration generation and scheduling for reconfigurable flow lines using non-dominated sorting genetic algorithm II (NSGA-II), which more objectives were concerned in their further study (Dou et al. 2020). Bortolini et al. (2021) have done a similar work adopting Bi-objective optimization for configuration design. Ameer and Dahane (2023) introduces a modular machine configuration approach to overcome the limitation of single-machine type selection in RMS research, integrating process planning and SRMT/MRMT selection through a bi-objective model, and optimizing multi-spindle machine utilization. Benderbal et al. (2017) investigated the multi-objective optimization approach for configuration design of RMS under unavailability constraints.

In conclusion, existing research on the configuration design of RMS/D-RMS primarily focuses on aspects such as scalability, convertibility, and delayed reconfiguration. Common approaches to achieving configuration design include configuration selection, evaluation, and optimization. Recently, optimization methods have gained popularity for addressing configuration design problems, which are generally formulated as multi-objective optimization problems. Therefore, a multi-objective optimization method is proposed to find the optimal dividing point of D-RMS to complete configuration design involving four objectives, including investment cost of machine tools, reconfiguration cost, reconfiguration time, and delayed reconfiguration. In addition, the proposed optimization model is solved by non-dominated sorting genetic algorithm-II (NSGA-II) (Deb et al. 2002; Goyal et al. 2012) due to its advantage on solving multi-objective integer programming problem.

The remainder of this chapter is organized as follows: Sect. 2 introduces the concept of D-RMS. Also, the dividing point definition and problem analysis are given in Sect. 2. Section 3 presents the proposed multi-objective optimization method for D-RMS configuration design by optimizing of the dividing point, which is solved by NSGA-II. Section 4 provides a case study to demonstrate the effectiveness of the proposed configuration design method of D-RMS. Section 5 concludes this chapter.

2 Concept Retrospect and Problem Analysis

2.1 D-RMS and Dividing Point

According to the definition of Huang et al. (2019), D-RMS focuses more on convertibility of manufacturing system than scalability, which can adapt to the demand fluctuation on product variety with relative low product volume. The core philosophy of D-RMS is postponing the reconfiguration activities to the latter stages as far as possible. Similar to the idea of scalability planning adopting more investment on machine tools/system structure in the study of Koren et al. (2017), a little bit more functions should be invested in the construction of manufacturing system to maintain partial production activities during reconfiguration in the former stages of D-RMS. However, it is necessary to optimize the additional functions investments to take full advantage of D-RMS.

A typical D-RMS configuration consists of two subsystems, including subsystem 1 and subsystem 2, as shown in Fig. 1. The subsystem 1 remains unchanged during reconfiguration to maintain production activities through more functions investments than exactly needed. The subsystem 2 provides exact functions when needed through reconfiguration to improve production efficiency. There is a position dividing subsystem 1 and subsystem 2, which is named as dividing point of D-RMS. The dividing point is similar to the decoupling point of delayed product differentiation (Ko and Jack Hu 2008) that is the inspiration source of D-RMS. Actually, the dividing point can be bound with the last stage of the subsystem 1 for computation convenience. There are three types of machine tools to be selected during configuration design of D-RMS, including dedicated machine tools, flexible machine tools, and reconfigurable machine tools. Dedicated machine tools and flexible machine tools are two options for the subsystem 1. Dedicated machine tools are adopted to realize one fixed function in the subsystem 1. Flexible machine tools can realize more than one function to eliminate the potential reconfiguration activities in the subsystem 1. Flexible machine tools are the main solution to provide more functions than exactly needed to maintain production activities of the subsystem 1 during reconfiguration. Reconfigurable machine tools are the core equipment of D-RMS to execute reconfiguration activities in the subsystem 2. Also, reconfigurable machine tools can provide more than one functions through reconfiguration. The different position of dividing point will affect the investment strategy of machine tools, which will finally impact

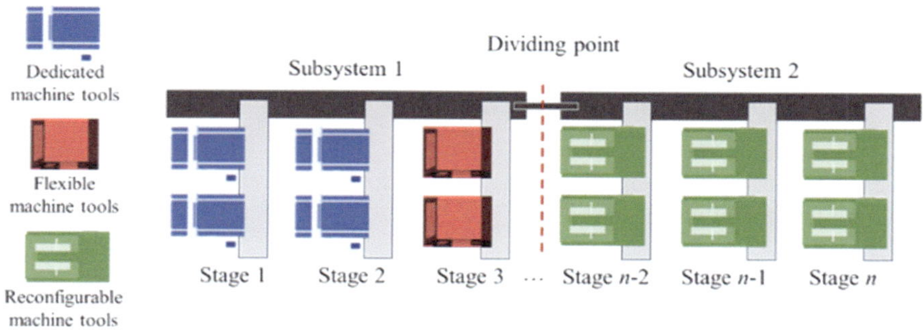

Fig. 1 Typical configuration of D-RMS

the production efficiency of D-RMS. It is necessary to optimize the position of dividing point during configuration design of D-RMS.

2.2 *Configuration Design Problem of D-RMS*

How to determine the optimal dividing point of D-RMS is the problem to be solved in this chapter. According to the definition of dividing point, the range of the subsystem 1 and the subsystem 2 will be changed when exploring the optimal position of dividing point. The different positions of dividing point results in different machine combinations among dedicated machine tools, flexible machine tools, and reconfigurable machine tools. The influence of different machine combinations should be concerned during dividing point optimization.

On the one hand, the dividing point should be postponed as far as possible to the latter stages according to the philosophy of D-RMS. The reconfiguration activities will be reduced. Reconfiguration effort is indicator of reconfiguration activities, including reconfiguration cost and time. However, the investment cost of reconfigurable machine tools is relatively inexpensive compared with flexible machine tools, which is more flexible than dedicated machine tools by providing customized functions. On the other hand, the production activities during reconfiguration benefits from additional function investment using flexible machine tools. Generally, the investment cost of flexible machine tools is rather more expensive than dedicated machine tools and reconfigurable machine tools. Therefore, the configuration design aims at seeking for balance between additional functions investment and reconfiguration effort by optimizing the dividing point of D-RMS configuration.

3 Optimization Model for Configuration Design of D-RMS

3.1 Assumptions and Nomenclature

To derive a simple yet insightful analytical model for configuration design of D-RMS, the following assumptions are made.

1. D-RMS is design and constructed around part family.
2. Only three types of machine tools can be adopted to design and construct a specific D-RMS, including dedicated machine tools, flexible machine tools and reconfigurable machine tools.
3. The investment cost of flexible machine tools is the most expensive. The investment cost of dedicated machine tools is the most inexpensive. The investment cost of reconfigurable machine tools is medium.
4. Only machine type in each stage of D-RMS configuration will be considered, whereas the machine numbers in each stage will not be concerned.
5. The part family information is already known, which should contain the characteristics of delayed reconfiguration. Without loss of generality, all parts from the specific part family share the same operation sequence length for simplicity.
6. Reconfiguration will be executed if the part types change (Only one part type is produced per time), where machine level reconfiguration is adopted in general.
7. The reconfiguration activities only happen in the subsystem 2 on the machine level.
8. For simplicity, all stages after dividing point of D-RMS are divided to subsystem 2.

To make the proposed optimization model more readable, the necessary explanations of associating parameters and variables are given in Table 1.

Table 1 Nomenclature

Term	Description
x	Decision variable, the position of dividing point
N	Total stage numbers of D-RMS
P	Total part numbers in a specific part family
M_n	Function numbers in n th stage
IC_{nm}	Basic investment cost of machine tools. That is, investment cost per function
$RC_{nmm'}$	Basic reconfiguration cost. That is, reconfiguration cost between two functions
$RT_{npp'}$	Reconfiguration time from part p to part p' in stage n

3.2 Mathematical Model

According to the definition of dividing point and the abovementioned assumptions, the configuration design process by optimizing the dividing point will make a balance among investment cost, reconfiguration cost, reconfiguration time, and delayed reconfiguration. The only decision variable x is integer. It is a typical multiple objectives integer programming problem. The corresponding mathematical models are shown in Eqs. (1), (2), (3) and (4), which attempt to minimize the investment cost, reconfiguration cost, and reconfiguration time and maximize delayed reconfiguration simultaneously. The constraint can be referred to Eq. (5).

$$\min \quad Investment\ cost = \sum_{n=1}^{x}\sum_{m=1}^{M_n} IC_{nm} + \sum_{n=x+1}^{N}\sum_{m=1}^{M_n/2} IC_{nm} \tag{1}$$

$$\min \quad Reconfiguration\ cost = \sum_{n=x+1}^{N}\sum_{m=1}^{M_n}\sum_{m'=m+1}^{M_n} RC_{nmm'} \tag{2}$$

$$\min \quad Reconfiguration\ time = \sum_{p=1}^{P}\sum_{p'=p+1}^{P} \max\left\{RT_{npp'} | n = x+1, x+2, \ldots N\right\} \tag{3}$$

$$\max \quad Delayed\ reconfiguration = x \tag{4}$$

$$1 < x < N \tag{5}$$

The first optimization objective referring to Eq. (1) concerns the influence on investment cost when changing the dividing point. The first item of Eq. (1) calculates the investment cost of machine tools in the subsystem 1, which consists of dedicated machine tools or flexible machine tools. Set the investment cost of dedicated machine tools as the basic investment cost of machine tools, that is, IC_{nm}. n is the index of stages. M_n is the index of the function number in stage n. $M_n = 1$ means a dedicated machine tools locating in stage n, however, $M_n > 1$ means a flexible machine tools locating in stage n. The value of M_n can be calculated according to the operation sequences of the specific part family. Similarly, the second item of Eq. (1) calculates the investment cost of machine tools in the subsystem 2 referring to reconfigurable machine tools. It is still calculated based on the basic investment cost, where $\lceil M_n/2 \rceil$ is divided by 2 and rounds up to an integer to show the medium investment cost of reconfigurable machine tools comparing with dedicated machine tools and flexible machine tools.

The second optimization objective referring to Eq. (2) addresses the reconfiguration cost problem during configuration design of D-RMS, which always happens in the stages of the subsystem 2 behind the dividing point. Considering that it is not possible to know specific reconfiguration demand during configuration design period, all reconfiguration activities happening in a specific stage in the subsystem 2 is counted during reconfiguration cost optimization. $RC_{nmm'}$ is the reconfiguration cost between two functions, which is adopted as basic reconfiguration cost. The reconfiguration cost of a specific stage is associated with function numbers referring to M_n.

The third optimization objective referring to Eq. (3) attempts to minimize the reconfiguration time during configuration design of D-RMS. Similar to reconfiguration cost calculation, all reconfiguration activities should be concerned during reconfiguration time optimization. However, the reconfiguration activities between two parts from the specific part family could be executed simultaneously. Therefore, although reconfiguration activities could happen in all the stage of the subsystem 2, only the maximum reconfiguration time will be counted when reconfiguring between two parts, that is $\max\{RT_{npp'}|n\}$.

The fourth optimization objective referring to Eq. (4) considers the characteristics of delayed reconfiguration during dividing point optimization, which is the only optimization objective pursuing maximum value. It is consistent with the D-RMS philosophy that tries to postpone the dividing point as latter as possible.

Equation (5) is the only constraint is the proposed optimization model for configuration design of D-RMS by optimizing dividing point. According to the inequality constraints in Eq. (5), the range of dividing point is limited within the configuration stages of D-RMS.

3.3 Solution by NSGA-III

The procedure of NSGA-II is shown as following:

Step 1. Initialization.
Initializing NSGA-II parameters and variables.
Step 2. Encoding and initial parent population generation.
According to the proposed optimization model, it is a sole decision variable problem. Considering that the sole decision variable x is integer, binary encoding is adopted. For example, if $x = 2$, the corresponding coding is [0 0 1 0]; if $x = 5$, the corresponding coding is [0 1 0 1]. Initially, a random parent population with Q individual is generated based on the binary encoding.
Step 3. Fast non-dominated sorting of parent population.
The four objective values of each individual in the initial population are calculated, which the dominant numbers of each individual are counted. The individual will be distributed to different fronts according to the dominant numbers. For each front, the individuals are sorted based on crowding distance.
Step 4. Jude the termination condition.
If the maximum generation is reached, the optimization process is complete, and the Pareto optimal solution will be obtained; otherwise, switch to step 5.

Step 5. Offspring population generation using elitism strategy.
Perform the selection, crossover, and mutation to generate an offspring population. To ensure the feasibility of selection result, the individual in the first front will be selected firstly, however, an individual will be eliminated if cannot satisfy the constraint in Eq. (5). Binary tournament selection is adopted to select Q individuals. And then, real code crossover, and polynomial mutation are executed on the selected Q individuals to generate the offspring population.
Step 6. Merge parent population and offspring population.
Merge the parent population and the offspring population to form a new population with size of $2Q$. Again, go back to step 3 for fast non-dominated sorting.

The flow chart of adopted NSGA-II is shown in Fig. 2.

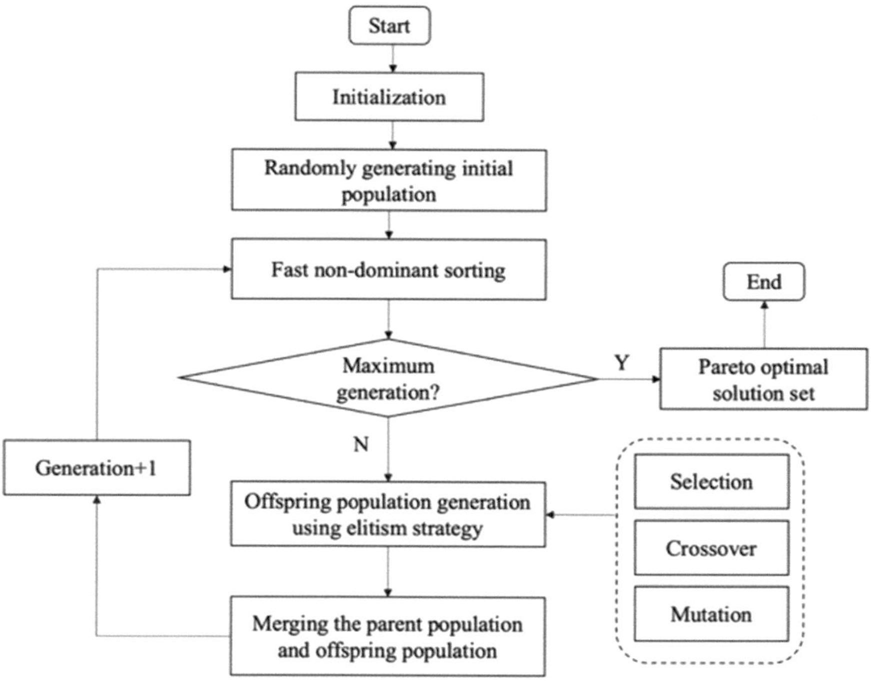

Fig. 2 The flow chart of elitist NSGA-II

4 Case Study

4.1 Configuration Design Based on Part Family 1

Part family 1 consists of three parts with perfect characteristics of delayed reconfiguration, as shown in Fig. 3. The corresponding operation sequences are part $a = \{1, 2, 3, 4, 5, 6\}$, part $b = \{1, 2, 3, 7, 8, 9\}$, and part $c = \{1, 2, 3, 10, 11, 12\}$. There is only one decoupling point among these three parts. It is obvious that the first three operations will remain unchanged during reconfiguration. The dividing point optimization process will focus on the influence of the last three operations. The dividing point could be the same to the decoupling point of the part family in this case. To explore the impact of investment cost of machine tools, reconfiguration cost and reconfiguration time, control variate method is adopted during experiments. The values of these parameters are given in Table 2. Due to the unit of these parameters will not affect the optimization result, this factor is not considered during experiments. To avoid unconscionable domination objective during optimization, the basic investment cost of machine tools, the basic reconfiguration cost, and the reconfiguration time should be preprocessed as similar magnitudes. Besides, the value of parameter P, M, and N are $P = 3$, $M = [1\ 1\ 1\ 3\ 3\ 3]$, and $N = 6$, respectively.

According the experiment design in Table 2, the first 4 experiments attempt to explore the impact of basic investment cost changes. The experiment results are shown in Fig. 4, including pareto front and solution details. There are four objectives for optimization resulting in high-dimensional graph during drawing pareto front. To make it more visual, reconfiguration time, reconfiguration cost, and investment cost are selected as the coordinates to present pareto front resulting in three-dimensional graph. The fourth objective, delayed reconfiguration (the same with decision variable x), is labeled near each point of pareto front. According to Fig. 4, there are three points in the pareto front of each experiment. Namely, there are three non-dominant solutions of each experiment. Take experiment 1 as example, in solution 1, investment

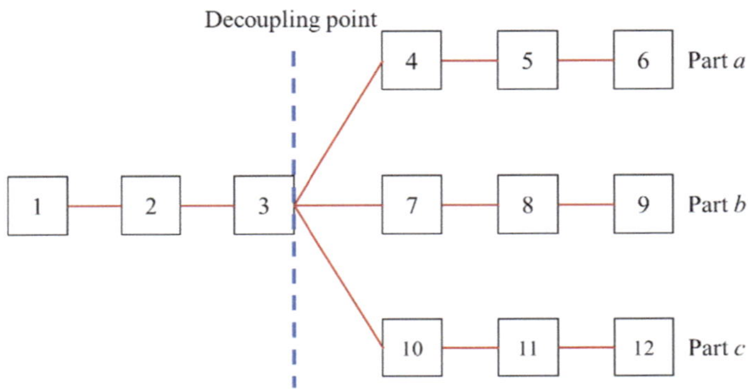

Fig. 3 Operation sequences of part family 1

Table 2 Experiment setting of part family 1

Experiment	Parameter	Value	Experiment	Parameter	Value
No. 1	*IC*	2	No. 2	*IC*	4
	RC	1		*RC*	1
	RT	$\begin{bmatrix} 1 & 1 & 1 \\ 1 & 1 & 1 \\ 1 & 1 & 1 \end{bmatrix}_{abc}$		*RT*	$\begin{bmatrix} 1 & 1 & 1 \\ 1 & 1 & 1 \\ 1 & 1 & 1 \end{bmatrix}_{abc}$
No. 3	*IC*	10	No. 4	*IC*	20
	RC	1		*RC*	1
	RT	$\begin{bmatrix} 1 & 1 & 1 \\ 1 & 1 & 1 \\ 1 & 1 & 1 \end{bmatrix}_{abc}$		*RT*	$\begin{bmatrix} 1 & 1 & 1 \\ 1 & 1 & 1 \\ 1 & 1 & 1 \end{bmatrix}_{abc}$
No. 5	*IC*	10	No. 6	*IC*	10
	RC	3		*RC*	6
	RT	$\begin{bmatrix} 1 & 1 & 1 \\ 1 & 1 & 1 \\ 1 & 1 & 1 \end{bmatrix}_{abc}$		*RT*	$\begin{bmatrix} 1 & 1 & 1 \\ 1 & 1 & 1 \\ 1 & 1 & 1 \end{bmatrix}_{abc}$
No. 7	*IC*	10	No. 8	*IC*	10
	RC	12		*RC*	3
	RT	$\begin{bmatrix} 1 & 1 & 1 \\ 1 & 1 & 1 \\ 1 & 1 & 1 \end{bmatrix}_{abc}$		*RT*	$\begin{bmatrix} 2 & 2 & 2 \\ 2 & 2 & 2 \\ 2 & 2 & 2 \end{bmatrix}_{abc}$
No. 9	*IC*	10	No. 10	*IC*	10
	RC	3		*RC*	3
	RT	$\begin{bmatrix} 4 & 4 & 4 \\ 4 & 4 & 4 \\ 4 & 4 & 4 \end{bmatrix}_{abc}$		*RT*	$\begin{bmatrix} 1 & 2 & 3 \\ 1 & 1 & 1 \\ 1 & 1 & 1 \end{bmatrix}_{abc}$
No. 11	*IC*	10	No. 12	*IC*	10
	RC	3		*RC*	3
	RT	$\begin{bmatrix} 3 & 2 & 1 \\ 1 & 1 & 1 \\ 1 & 1 & 1 \end{bmatrix}_{abc}$		*RT*	$\begin{bmatrix} 1 & 1 & 1 \\ 2 & 1 & 1 \\ 3 & 1 & 1 \end{bmatrix}_{abc}$
No. 13	*IC*	10	No. 14	*IC*	10

(continued)

Table 2 (continued)

Experiment	Parameter	Value	Experiment	Parameter	Value
	RC	3		*RC*	3
	RT	$\begin{bmatrix} 1 & 1 & 1 \\ 1 & 1 & 2 \\ 1 & 1 & 3 \end{bmatrix}_{abc}$		*RT*	$\begin{bmatrix} 1 & 2 & 1 \\ 3 & 1 & 2 \\ 2 & 4 & 3 \end{bmatrix}_{abc}$

cost = 18, reconfiguration cost = 9, reconfiguration time = 3, delayed reconfiguration = 3. The first three operations of part family 1 will be distributed to the subsystem 1 of D-RMS. The last three operations of part family 2 will be distributed to the subsystem 2. The first three operations of these parts in part family 1 shares the same functions, which will remain unchanged during reconfiguration. The last three operations of these parts in part family 1 require quite different functions, which can be realized through reconfiguration. Therefore, solution 1 is consistent with intuition. In solution 2, investment cost = 20, reconfiguration cost = 6, reconfiguration time = 3, delayed reconfiguration = 4. Compared with solution 1, the investment cost increases from 18 to 20 as the decision variable is postponed to next stage, whereas the reconfiguration cost decreases from 9 to 6. Besides, the delayed reconfiguration increases from 3 to 4 (higher is better). So, solution 2 is reasonable as well. In solution 3, investment cost = 22, reconfiguration cost = 3, reconfiguration time = 3, delayed reconfiguration = 5, which is similar to solution 2.

It is difficult to make distinguish among these three solutions in current conditions. Generally, reasonable weights can be assigned to the proposed four objectives to create an index for further comparison among these solutions, as shown in Eq. (6).

$$\begin{aligned} F = {} & \omega_1 \textit{ investment cost} + \omega_2 \textit{ reconfiguration cost} + \omega_3 \textit{ reconfiguration time} \\ & - \omega_4 \textit{ delayed reconfiguration} \end{aligned} \tag{6}$$

where F is the comparison index to find the solution with lowest value. ω_1, ω_2, ω_3, and ω_4 are the weight assigning to the proposed four objectives respectively and $\omega_1 + \omega_2 + \omega_3 + \omega_4 = 1$. For comparison convenience, these weights are set as the same in case study, that is, $\omega_1 = \omega_2 = \omega_3 = \omega_4 = 0.25$.

Take experiment 1 as example again, the comparison index values of these three solutions are $F_{s1} = 6.75$, $F_{s2} = 6.25$, and $F_{s3} = 6.75$, respectively. In this condition, solution 2 is the best one with lowest comparison index value. The configuration of solution 2 in experiment 1 is shown in Fig. 5, where the dividing point is bound to stage 4 and a flexible machine tools is selected to achieve function 4, 7, and 10 simultaneously in the subsystem 1. Obviously, the best solution could be changed under different weight combinations, which depends on the preference of decision maker and specific conditions.

According to the results of the first 4 experiments, the values of reconfiguration cost, reconfiguration time, and delayed reconfiguration remain unchanged under

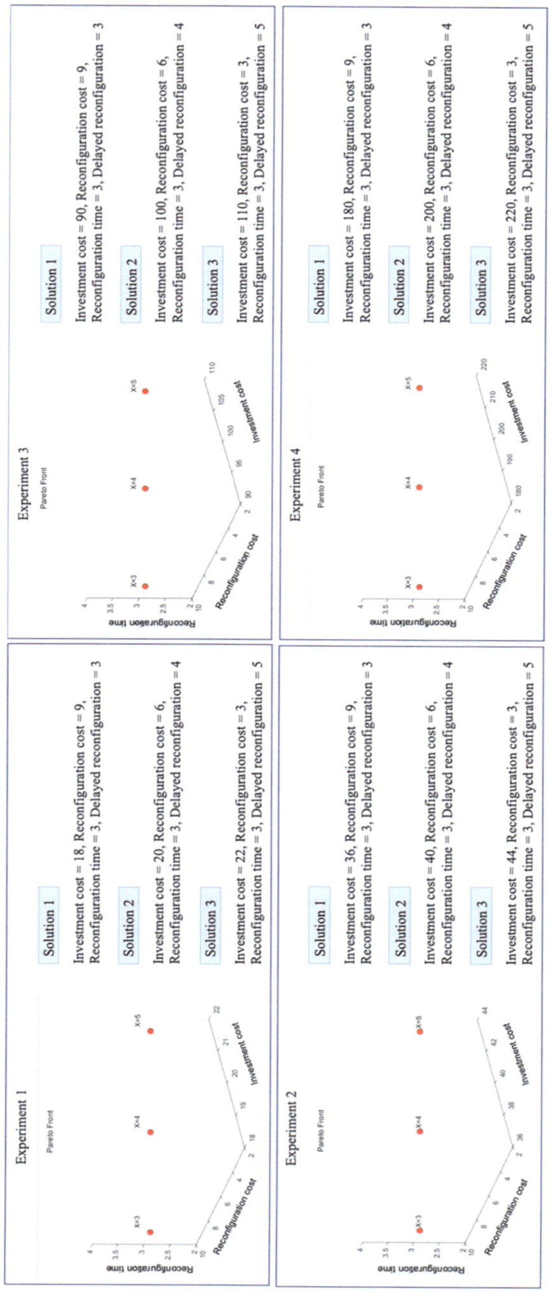

Fig. 4 The results of experiment 1, 2, 3, and 4

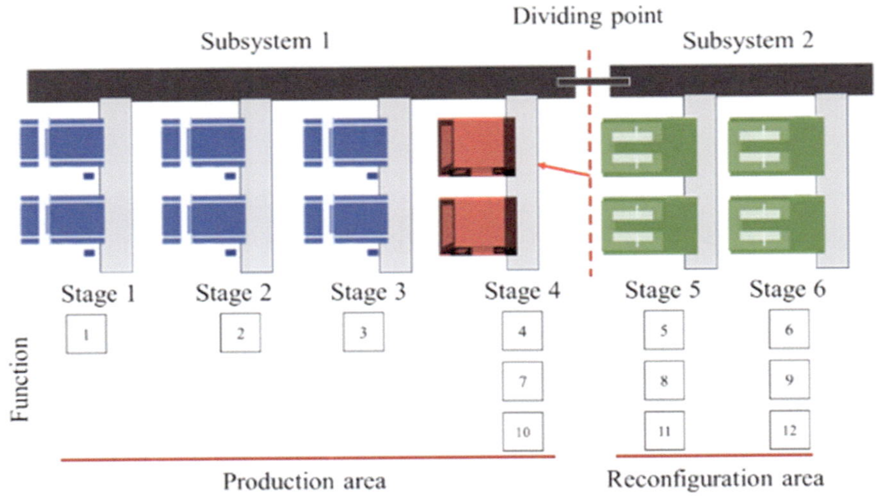

Fig. 5 Configuration of solution 2 in experiment 1 based on part family 1

different basic investment cost situations. However, the investment cost increases linearly with basic investment cost. The investment cost could dominate other objectives, which it is obstacle to postpone the dividing point to latter stage blocking by extremely high investment cost. Take experiment 4 as example, the investment cost of solution 1, solution 2, and solution 3 are 180, 200, and 220 respectively, which is far more than the value of experiment 1. The corresponding comparison index values of these three solutions are $F_{s1} = 47.25$, $F_{s2} = 51.25$, and $F_{s3} = 56.25$ respectively. Here, solution 1 is the best one with dividing point = 3. The configuration of solution 1 in experiment 4 is shown in Fig. 6. No flexible machine tools is selected to realize multiple functions that are distributed to the subsystem 2 for reconfiguration.

Experiment 5, 6, and 7 are executed to analyze the influence of basic reconfiguration cost changes, where the parameter settings are the same to experiment 3 excepting basic reconfiguration, as shown in Fig. 7. Similarly, the reconfiguration cost of each experiment increases linearly with basic reconfiguration cost. In experiment 3, the comparison index values are $F_{s1} = 24.75$, $F_{s2} = 26.25$, and $F_{s3} = 27.75$, respectively. In experiment 5, the comparison index values are $F_{s1} = F_{s2} = F_{s3} = 29.25$. In experiment 6, the comparison index values are $F_{s1} = 36$, $F_{s2} = 33.75$, and $F_{s3} = 31.5$, respectively. In experiment 7, the comparison index values are $F_{s1} = 49.5$, $F_{s2} = 42.75$, and $F_{s3} = 36$, respectively. The increase of basic reconfiguration cost reverses the optimization preference by comparing experiment 5, 6, 7 with experiment 3. The comparison index values of experiment 3 increase progressively from solution 1 to solution 3, whereas they are progressive decrease in experiment 6 and 7. It can be concluded that the solution 3 of these three experiments (experiment 5, 6, and 7) is the best one respectively,

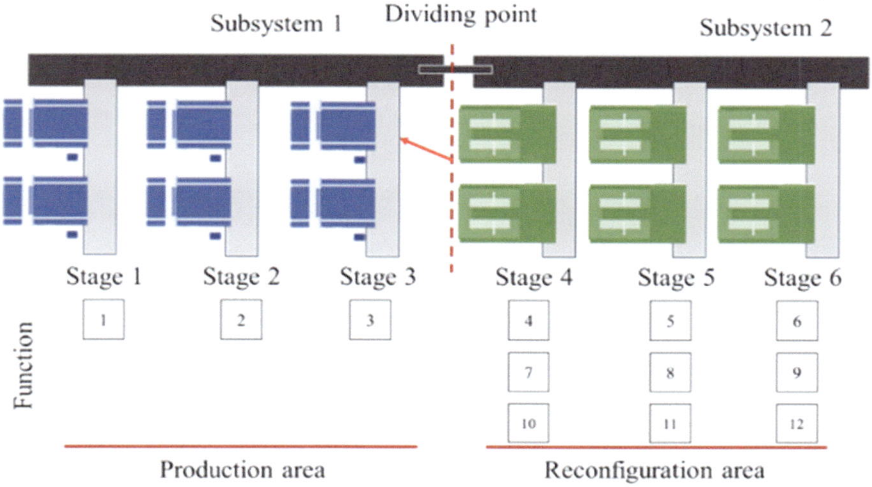

Fig. 6 Configuration of solution 1 in experiment 4 based on part family 1

which means that the proposed method tries to decrease the reconfiguration activities during optimization due to the increase of basic reconfiguration cost.

The experiment results of reconfiguration time changes are shown in Fig. 8. The characteristics of reconfiguration time is different with investment cost and reconfiguration time. Due to reconfiguration activities can be executed simultaneously, only the largest reconfiguration time will affect the optimization results. According to the results of experiment 8 and 9, the comparison index values are $F_{s1} = F_{s2} = F_{s3} = 30$ and $F_{s1} = F_{s2} = F_{s3} = 31.5$ respectively, which share the same trend with the changes of basic investment cost and basic reconfiguration cost. However, the dividing point optimization is sensitive to the position of largest reconfiguration time. Take experiment 10 and 11 as example, the largest reconfiguration times appear in 6th operation and 4th operation respectively. The comparison index values of experiment 10 are $F_{s1} = F_{s2} = F_{s3} = 29.75$, however, experiment 11 are $F_{s1} = 29.75$, $F_{s2} = 29.5$, and $F_{s3} = 29.25$, respectively. Due to the largest reconfiguration time of experiment 10 appears in the last operation, the corresponding comparison index values are the same when the value of decision variable x changes from 4th operation to 6th operation. The comparison index values of experiment 11 decrease from 29.75 to 29.25 because of the largest reconfiguration time changes from 3 to 1 when decision variable x changes from 4th operation to 6th operation. The experiment 12 shares the same characteristics with experiment 11. Similarly, the experiment 13 shares the same characteristics with experiment 10.

The experiment 14 are a comprehensive example to explore the influence of reconfiguration time changes with complex reconfiguration time combination. The comparison index values of experiment 14 are $F_{s1} = 30.75$, $F_{s2} = 30.5$, and $F_{s3} = 30$

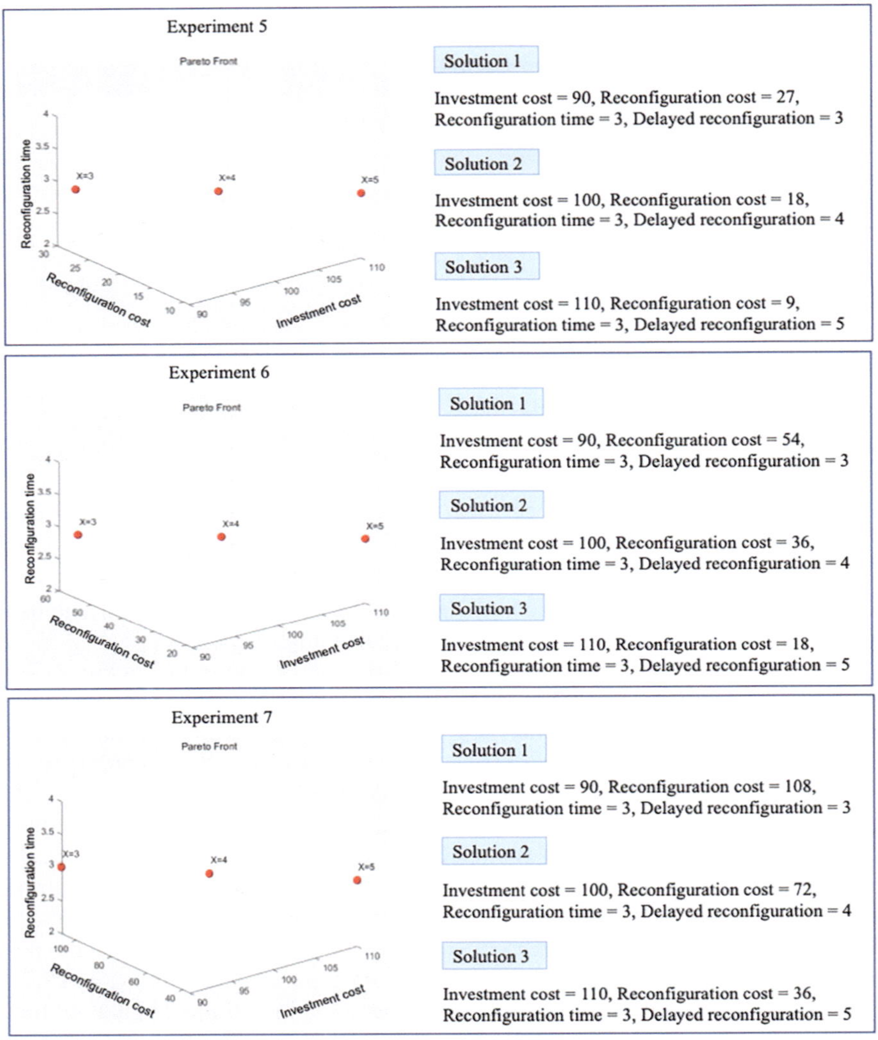

Fig. 7 The results of experiment 5, 6, and 7

respectively. The solution 3 in experiment 14 is the best one, which the corresponding configuration is shown in Fig. 9.

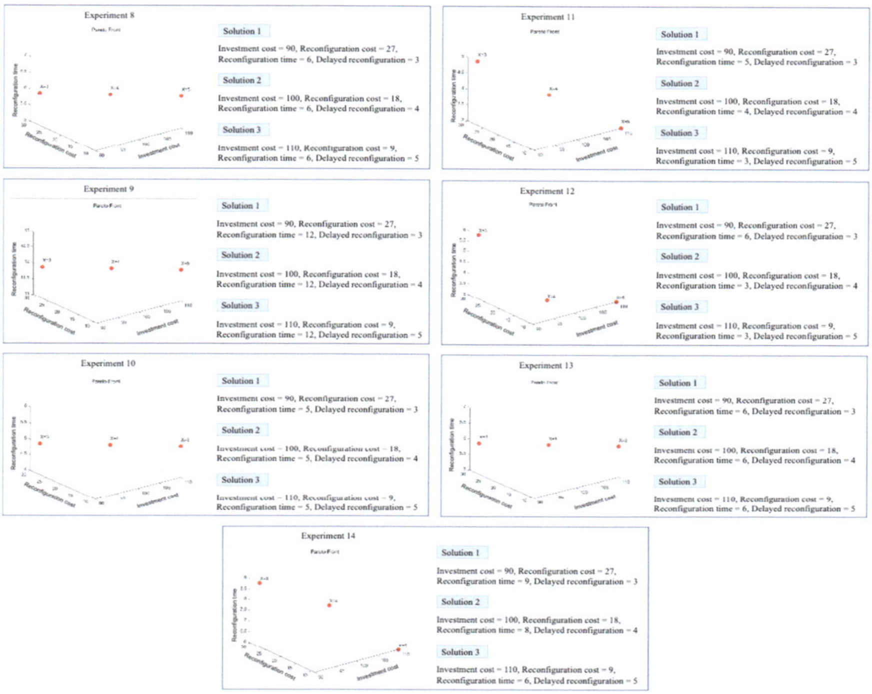

Fig. 8 Experiment results of reconfiguration time changes

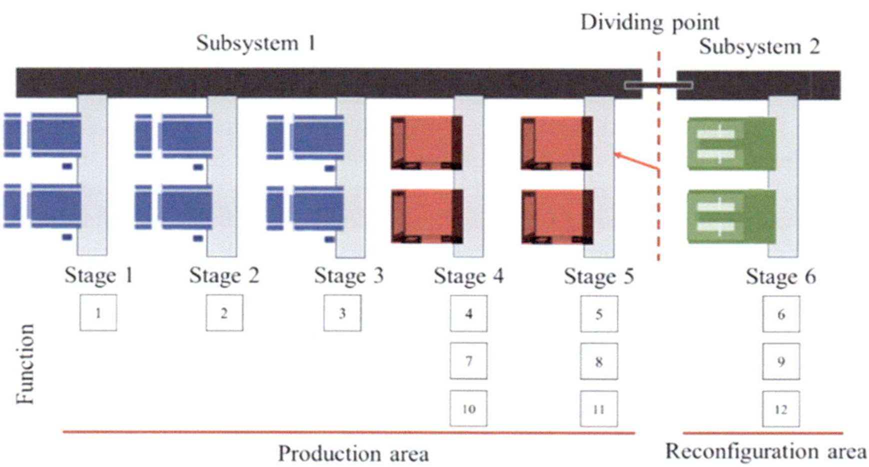

Fig. 9 Configuration of solution 3 in experiment 4 based on part family 1

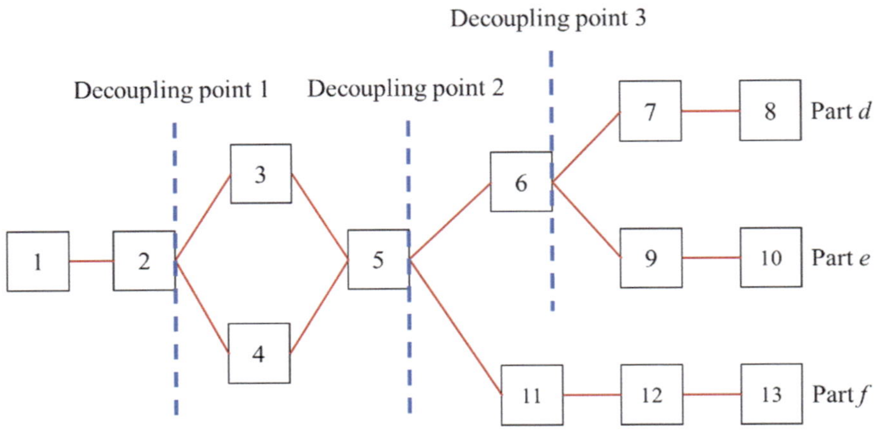

Fig. 10 The operation sequences of part family 2

4.2 *Configuration Design Based on Part Family 2*

Similarly, part family 2 also consists of three parts as shown in Fig. 10, that is, part $d = \{1, 2, 3, 5, 6, 7, 8\}$, part $e = \{1, 2, 4, 5, 6, 9, 10\}$, and part $f = \{1, 2, 3, 5, 11, 12, 13\}$. However, the operation details of these three parts are far more complex than part family 1, which more than one decoupling point among these three parts. The settings of basic investment cost, basic reconfiguration cost, and reconfiguration time are shown in Table 3. Similarly, the value of parameter *P*, *M*, and *N* are $P = 3$, $M = [1\ 1\ 2\ 1\ 2\ 3\ 3]$, and $N = 7$, respectively.

The experiment results of configuration design based on part family 2 are shown in Fig. 11. There are two non-dominant solutions of each experiment with dividing point in 5th operation or 6th operation separately. The previous proposed comparison index with the same weight assignment is adopted here again to make further analysis, which the computation results are given in Table 4. According to table 4, although decoupling point 1 of part family 2 appears earlier than part family 1, the dividing point optimization still prefers latter stages in general. Take experiment 1 in Fig. 11 as example, the comparison index values of the two solutions are 6.5 and 6 respectively. Namely, solution 2 in experiment 1 is the best one. The corresponding configuration of the solution 2 is shown in Fig. 12. However, the optimal solution could reverse if basic investment cost increase according to the comparison index values of experiment 1 and experiment 2, which the optimal solution of experiment 2 changes to solution 1. The increase of basic reconfiguration cost can draw the similar conclusion according to the comparison index values of experiment 3 and experiment 4. Also, the increase of reconfiguration time could reduce the preference of solution with latter dividing point according to the experiment results of experiment 3 and experiment 7, whereas the maximum reconfiguration time appears before 5th operation or last operation will not change the optimization results according to

Table 3 Experiment setting of part family 2

Experiment	Parameter	Value	Experiment	Parameter	Value
No. 1	IC	2	No. 2	IC	10
	RC	1		RC	1
	RT	$\begin{bmatrix} 1\,0\,0 & 1\,1 \\ 0\,0\,1 & 1\,1 \\ 1\,0\,1 & 1\,1 \end{bmatrix}_{def}$		RT	$\begin{bmatrix} 1\,0\,0 & 1\,1 \\ 0\,0\,1 & 1\,1 \\ 1\,0\,1 & 1\,1 \end{bmatrix}_{def}$
No. 3	IC	10	No. 4	IC	10
	RC	3		RC	12
	RT	$\begin{bmatrix} 1\,0\,0 & 1\,1 \\ 0\,0\,1 & 1\,1 \\ 1\,0\,1 & 1\,1 \end{bmatrix}_{def}$		RT	$\begin{bmatrix} 1\,0\,0 & 1\,1 \\ 0\,0\,1 & 1\,1 \\ 1\,0\,1 & 1\,1 \end{bmatrix}_{def}$
No. 5	IC	10	No. 6	IC	10
	RC	3		RC	3
	RT	$\begin{bmatrix} 1\,0\,0 & 1\,1 \\ 0\,0\,2 & 1\,1 \\ 1\,0\,2 & 1\,1 \end{bmatrix}_{def}$		RT	$\begin{bmatrix} 6\,0\,0 & 1\,1 \\ 0\,0\,1 & 1\,1 \\ 6\,0\,1 & 1\,1 \end{bmatrix}_{def}$
No. 7	IC	10	No. 8	IC	10
	RC	3		RC	3
	RT	$\begin{bmatrix} 1\,0\,0 & 4\,1 \\ 0\,0\,1 & 1\,1 \\ 1\,0\,1 & 1\,1 \end{bmatrix}_{def}$		RT	$\begin{bmatrix} 1\,0\,0 & 1\,4 \\ 0\,0\,1 & 1\,1 \\ 1\,0\,1 & 1\,1 \end{bmatrix}_{def}$
No. 9	IC	10	No. 10	IC	10
	RC	3		RC	3
	RT	$\begin{bmatrix} 4\,0\,0 & 1\,1 \\ 0\,0\,3 & 1\,1 \\ 4\,0\,3 & 1\,1 \end{bmatrix}_{def}$		RT	$\begin{bmatrix} 1\,0\,0 & 1\,4 \\ 0\,0\,3 & 1\,1 \\ 1\,0\,3 & 1\,1 \end{bmatrix}_{def}$

experiment 5, 6, 8, 9, and 10 (solution 1 and solution 2 share the same comparison index value in these experiments).

According to the results of previous two case studies, the implementation processes of the proposed configuration design of D-RMS by optimizing the dividing point is presented. The experiment results show that the proposed method can realize the philosophy of D-RMS with the point of pareto front locating at the latter stages of D-RMS. Moreover, the influences of investment cost, reconfiguration cost, and reconfiguration time are analyzed in these two case studies. Lower basic investment

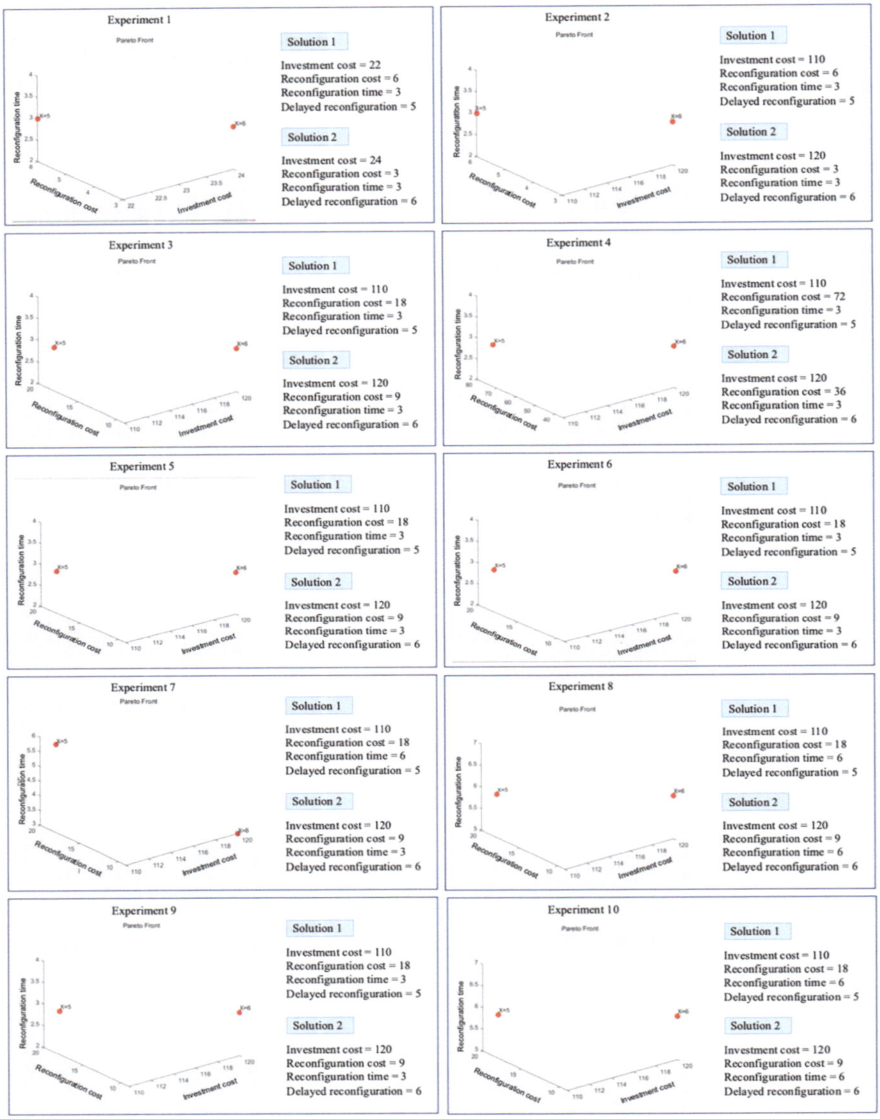

Fig. 11 Experiment results of part family 2

cost or reconfiguration cost will postpone the dividing point to latter stages of D-RMS. although reconfiguration time shares the similar results, the optimization result is sensitive to the operation position of maximum reconfiguration time. Above all, these two case studies demonstrate the effectiveness of the proposed configuration design method of D-RMS, which can be used to guide the construction of D-RMS.

Table 4 Comparison index values of the part family 2

Experiment	Solution	Comparison index	Experiment	Solution	Comparison index
No. 1	1	6.5	No. 2	1	28.5
	2	6		2	30
No. 3	1	31.5	No. 4	1	45
	2	31.5		2	38.25
No. 5	1	31.5	No. 6	1	31.5
	2	31.5		2	31.5
No. 7	1	32.25	No. 8	1	32.25
	2	31.5		2	32.25
No. 9	1	31.5	No. 10	1	32.25
	2	31.5		2	32.25

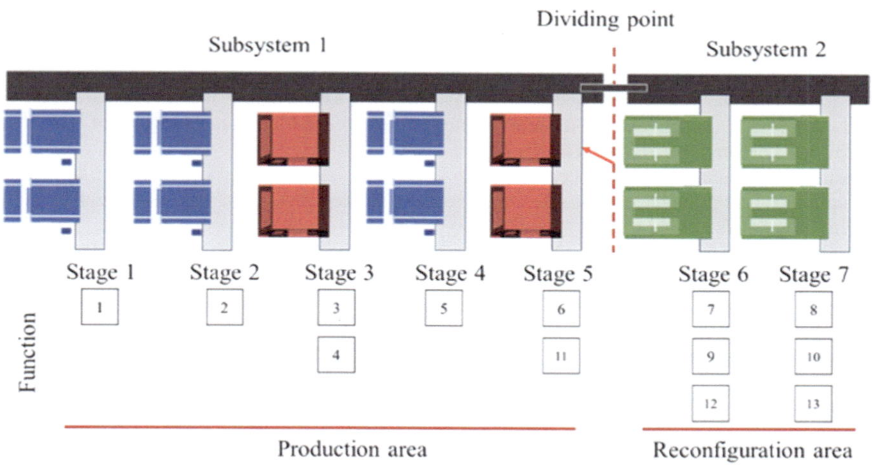

Fig. 12 Configuration of solution 2 in experiment 1 based on part family 2

5 Conclusion

In this chapter, the configuration design problem of D-RMS is addressed based on NSGA-II. A multiple objectives optimization model is constructed with the consideration of investment cost, reconfiguration cost, reconfiguration time, and delayed reconfiguration. NSGA-II is adopted to solve this integer optimization model. And then, two cases are presented to implement the proposed method for configuration design of D-RMS based two part families with different characteristics. The experiment results show that the proposed optimization model is able to postpone the dividing point to the latter stages of manufacturing system, which is consistent with the philosophy of D-RMS. Also, the dividing point could be adjusted slightly among latter stages if the basic investment cost, basic reconfiguration cost, or reconfiguration time is changing.

References

Ameer M, Dahane M (2023) NSGA-III-based multi-objective approach for reconfigurable manufacturing system design considering single-spindle and multi-spindle modular reconfigurable machines. Int J Adv Manuf Technol 128(5–6):2499–2524

Andersen AL, Brunoe TD, Nielsen K, Rösiö C (2017) Towards a generic design method for reconfigurable manufacturing systems: analysis and synthesis of current design methods and evaluation of supportive tools. J Manuf Syst 42:179–195

Ashraf M, Hasan F (2018) Configuration selection for a reconfigurable manufacturing flow line involving part production with operation constraints. Int J Adv Manuf Technol 98(5):2137–2156

Battaïa O, Dolgui A, Guschinsky N (2017) Decision support for design of reconfigurable rotary machining systems for family part production. Int J Prod Res 55(5):1368–1385

Benderbal HH, Dahane M, Benyoucef L (2017) Flexibility-based multi-objective approach for machines selection in reconfigurable manufacturing system (RMS) design under unavailability constraints. Int J Prod Res 55(20):6033–6051

Benderbal HH, Dahane M, Benyoucef L (2018) Modularity assessment in reconfigurable manufacturing system (RMS) design: an archived multi-objective simulated annealing-based approach. Int J Adv Manuf Technol 94(1):729–749

Bensmaine A, Dahane M, Benyoucef L (2013) A non-dominated sorting genetic algorithm based approach for optimal machines selection in reconfigurable manufacturing environment. Comput Ind Eng 66(3):519–524

Bortolini M, Galizia FG, Mora C (2018) Reconfigurable manufacturing systems: literature review and research trend. J Manuf Syst 49:93–106

Bortolini M, Botti L, Galizia FG, Regattieri A (2021) Bi-objective design and management of reconfigurable manufacturing systems to optimize technical and ergonomic performances. Appl Sci 11(1):263

Cerqueus A, Delorme X (2023) Evaluating the scalability of reconfigurable manufacturing systems at the design phase. Int J Prod Res 61(23):8080–8093

Deb K, Pratap A, Agarwal S, Meyarivan TAMT (2002) A fast and elitist multiobjective genetic algorithm: NSGA-II. IEEE Trans Evol Comput 6(2):182–197

Dou J, Li J, Su C (2016) Bi-objective optimization of integrating configuration generation and scheduling for reconfigurable flow lines using NSGA-II. Int J Adv Manuf Technol 86(5):1945–1962

Dou J, Li J, Xia D, Zhao X (2020) A multi-objective particle swarm optimisation for integrated configuration design and scheduling in reconfigurable manufacturing system. Int J Prod Res 1–21

ElMaraghy HA (2005) Flexible and reconfigurable manufacturing systems paradigms. Int J Flex Manuf Syst 17(4):261–276

Fang W, Guo Y, Huang S, Liu D, Cui S, Liao W, Hong D (2021) Big data driven intelligent production control of discrete manufacturing process. J Mech Eng 57(20):277–291

Gauss L, Lacerda DP, Sellitto MA (2019) Module-based machinery design: a method to support the design of modular machine families for reconfigurable manufacturing systems. Int J Adv Manuf Technol 102(9):3911–3936

Goyal KK, Jain PK, Jain M (2012) Optimal configuration selection for reconfigurable manufacturing system using NSGA II and TOPSIS. Int J Prod Res 50(15):4175–4191

Huang S, Yan Y (2019) Part family grouping method for reconfigurable manufacturing system considering process time and capacity demand. Flex Serv Manuf J 31(2):424–445

Huang S, Yan Y (2020) Design of delayed reconfigurable manufacturing system based on part family grouping and machine selection. Int J Prod Res 58(14):4471–4488
Huang S, Wang G, Yan Y (2019) Delayed reconfigurable manufacturing system. Int J Prod Res 57(8):2372–2391
Ko J, Jack Hu S (2008) Balancing of manufacturing systems with complex configurations for delayed product differentiation. Int J Prod Res 46(15):4285–4308
Koren Y (2013) The rapid responsiveness of RMS. Int J Prod Res 51(23–24):6817–6827
Koren Y, Wang W, Gu X (2017) Value creation through design for scalability of re configurable manufacturing systems. Int J Prod Res 55(5):1227–1242
Koren Y, Gu X, Guo W (2018a) Reconfigurable manufacturing systems: principles, design, and future trends. Front Mech Eng 13(2):121–136
Koren Y, Gu X, Guo W (2018b) Choosing the system configuration for high-volume manufacturing. Int J Prod Res 56(1–2):476–490
Kusiak A (2018) Smart manufacturing. Int J Prod Res 56(1–2):508–517
Lasi H, Fettke P, Kemper HG, Feld T, Hoffmann M (2014) Industry 4.0. Bus & Inf Syst Eng 6(4):239–242
Lu Y, Liu C, Kevin I, Wang K, Huang H, Xu X (2020) Digital Twin-driven smart manufacturing: connotation, reference model, applications and research issues. Robot Comput-Integr Manuf 61:101837
Mansour H, Afefy IH, Taha SM (2023) Simultaneous layout design optimization with the scalable reconfigurable manufacturing system. Prod Eng Res Devel 17(3):565–573
Meng F, Song P (2022) Influencing mechanism of intelligent manufacturing ecosystem on intelligent transition of manufacturing enterprises. Sci Res Manag 43(4):37
Moghaddam SK, Houshmand M, Fatahi Valilai O (2018) Configuration design in scalable reconfigurable manufacturing systems (RMS); a case of single-product flow line (SPFL). Int J Prod Res 56(11):3932–3954
Moghaddam SK, Houshmand M, Saitou K, Fatahi Valilai O (2020) Configuration design of scalable reconfigurable manufacturing systems for part family. Int J Prod Res 58(10):2974–2996
Nti IK, Adekoya AF, Weyori BA, Nyarko-Boateng O (2021) Applications of artificial intelligence in engineering and manufacturing: a systematic review. J Intell Manuf 1–21
Pal Singh P, Madan J, Singh H (2020) Composite performance metric for product flow configuration selection of reconfigurable manufacturing system (RMS). Int J Prod Res 1–21
Puik E, Telgen D, van Moergestel L, Ceglarek D (2017) Assessment of reconfiguration schemes for reconfigurable manufacturing systems based on resources and lead time. Robot Comput-Integr Manuf 43:30–38
Rezaee P, Moghaddam SK (2025) Assessment of design approaches for reconfigurable manufacturing systems based on forecasted demand data. Comput & Ind Eng 110878
Shang X, Milisavljevic-Syed J, Huang S, Wang G, Allen JK, Mistree F (2021) A key feature-based method for the configuration design of a reconfigurable inspection system. Int J Prod Res 59(9):2611–2623
Sigov A, Ratkin L, Ivanov LA, Xu LD (2022) Emerging enabling technologies for Industry 4.0 and beyond. Inf Syst Front 1–11
Singh A, Gupta S, Asjad M, Gupta P (2017) Reconfigurable manufacturing systems: journey and the road ahead. Int J Syst Assur Eng Manag 8(2):1849–1857
Tao F, Qi Q (2017) New IT driven service-oriented smart manufacturing: framework and characteristics. IEEE Trans Syst Man Cybern Syst 49(1):81–91
Tao F, Qi Q, Liu A, Kusiak A (2018) Data-driven smart manufacturing. J Manuf Syst 48:157–169
Wang GX, Huang SH, Yan Y, Du JJ (2017) Reconfiguration schemes evaluation based on preference ranking of key characteristics of reconfigurable manufacturing systems. The International Journal of Advanced Manufacturing Technology 89(5):2231–2249
Wang Y, Wang C, Xu Y, Sun R, Xiao K, Wang K (2024) Digital Twin Task Scheduling Method for Jobs of Intelligent Manufacturing Unit under Edge-cloud Collaboration. Journal of Mechanical Engineering 60(6):137–152

Yelles-Chaouche AR, Gurevsky E, Brahimi N, Dolgui A (2020) Reconfigurable manufacturing systems from an optimisation perspective: a focused review of literature. Int J Prod Res 1–19
Yu Z, Gong Y (2024) ChatGPT, AI-generated content, and engineering management. Front Eng Manag 11(1):159–166
Zhang J, Söpper M, Holzapfel F (2021) Attainable moment set optimization to support configuration design: a required moment set based approach. Appl Sci 11(8):3685
Zhang T, Homri L, Dantan JY, Siadat A (2023) Models for reliability assessment of reconfigurable manufacturing system regarding configuration orders. Reliab Eng Syst Saf 231:109035
Zhong RY, Xu X, Klotz E, Newman ST (2017) Intelligent manufacturing in the context of Industry 4.0: a review. Engineering 3(5):616–630
Zhou Z, Tao B (2023) Scientific System and Technology Framework of Digital Manufacturing. Journal of Mechanical Engineering 59(19):126–151
Zhou J, Li P, Zhou Y, Wang B, Zang J, Meng L (2018) Toward new-generation intelligent manufacturing. Engineering 4(1):11–20
Zhou G, Zhang C, Li Z, Ding K, Wang C (2020) Knowledge-driven digital twin manufacturing cell towards intelligent manufacturing. Int J Prod Res 58(4):1034–1051

Online Optimization of Reconfiguration Planning for SRMS Based on DQN

Sihan Huang, Zhicheng Peng, Jintang Huang, Guoxin Wang, and Yan Yan

Abstract Reconfigurable machine tools (RMTs) as key equipment of smart reconfigurable manufacturing systems (SRMS) can promote its flexibility when demand changes. The fundamental problem lies in dynamically reconfiguring the RMTs in SRMS efficiently and accurately by considering the flexibility of production precedence and operation sequences simultaneously. Therefore, a dynamic reconfiguration planning method for SRMS based on deep reinforcement learning is proposed in this chapter. The reconfiguration processes of SRMS are modeled by considering reconfiguration cost, moving cost, and processing cost. Deep Q-network (DQN) is adopted to find the optimal reconfiguration scheme with the highest return. A case study is presented to demonstrate the effectiveness and efficiency of the proposed method.

1 Introduction

Smart manufacturing systems (SMS) emerging in Industry 4.0 era are designed to respond in real-time to changing demands and conditions in factories and supply networks while meeting varying customer needs (Kusiak 2018; Zhenget al. 2018; Ahmed et al. 2022). Reconfigurable machine tools (RMT) play a fundamental role in enabling the changeability and reconfigurability of SMS, which are critical for ensuring responsiveness to demand fluctuations (Yin et al. 2012; Huang et al. 2022; Dhupia et al. 2007). SMS with RMT (named as SRMS) equipped with high flexibility are a new paradigm to deal with demand fluctuation growing out of individual customization. Additionally, digital twin as the synthesis of the emerging ICT and AI

S. Huang (✉) · Z. Peng · G. Wang · Y. Yan
School of Mechanical Engineering, Beijing Institute of Technology, Beijing, China
e-mail: hsh@bit.edu.cn

S. Huang · G. Wang · Y. Yan
Key Laboratory of Industry Knowledge and Data Fusion Technology and Application, Ministry of Industry and Information Technology, Beijing Institute of Technology, Beijing, China

J. Huang
BEIJING FANUC, Beijing, China

S. Huang et al. (eds.), *Design and Operation of Smart Reconfigurable Manufacturing Systems in Industry 4.0/5.0*, Springer Series in Advanced Manufacturing,
https://doi.org/10.1007/978-3-032-00284-6_4

is the core enabler of SRMS to increase responsiveness. Digital twin-driven SRMS are gradually unfolding the potential of accelerating the development of current industrial evolution.

The effective integration of emerging technologies to enhance reconfiguration efficiency and precision constitutes a critical challenge in implementing RMTs-enabled SMS. To address this challenge, this study proposes a dynamic reconfiguration planning method for digital twin-driven SRMS based on deep reinforcement learning to autonomously adapt and optimize production processes under fluctuating demand. The methodology formalizes RMT reconfiguration processes within a Markov decision process (MDP) framework, employing Deep Q-Network (DQN) to iteratively identify optimal reconfiguration strategies.

This chapter is structured as follows: Section 1 introduces the research background, industrial challenges, and innovation motivations in reconfigurable manufacturing systems. Section 2 systematically reviews existing literature on smart manufacturing reconfiguration, smart manufacturing reconfiguration driven by digital twin, and reinforcement learning applications, identifying critical research gaps. Section 3 formally defines the problem space through mathematical modeling of dynamic reconfiguration planning, emphasizing operational constraints and optimization objectives specific to reconfigurable machine tools (RMTs). Section 4 develops a novel deep reinforcement learning framework integrating DQN architecture, detailing the state-action-reward modeling paradigm and policy optimization mechanisms. Section 5 implements this framework in an automotive production case study, quantitatively validating performance metrics including reconfiguration efficiency and resource utilization rates. The concluding section discusses theoretical contributions to adaptive manufacturing.

2 Literature Review

The concept of reconfigurable manufacturing system (RMS) to explore reconfiguration in manufacturing systems was proposed by Koren et al. (1999), where RMT is the fundamental equipment of RMS to promote its responsiveness (Koren 2013). RMS is a promising manufacturing paradigm aiming at providing exactly functionality and capacity needed and exactly when needed (Huang et al. 2019). SMS with reconfigurability are the evolved version of RMS to some extent in Industry 4.0 era (Morgan et al. 2021; Tang et al. 2019; Wan et al. 2018), moving towards smart reconfigurable manufacturing systems. Lee and Ryu (2022) proposed a smart, self-reconfigurable manufacturing system and discussed its architecture and key features. Reconfiguration planning, i.e., how to reconfigure a manufacturing system is the key problem to be solved when demand changes. Reconfiguration planning is the key phase to successfully implementing RMS or SMS with reconfigurability. Many researchers investigated reconfiguration planning. Typically, this problem has been studied from two main perspectives, including scalability and convertibility (Bortolini et al. 2018; Yelles-Chaouche et al. 2021). Koren et al. (2017) studied the scalability planning problem of RMS to maximize system throughput after reconfiguration. Also, Koren et al. (2018) discussed the principles, design, and development trends of RMS. Ma et al. (2020) proposed a bilevel coordinated optimization model solved by a nesting genetic algorithm to address the reconfigurable process planning problem. Azab and ElMaraghy (2007) put forward mathematical modeling for reconfigurable process planning, in which a process plan reconfiguration index that captures the extent of changes in the plan and their implications has been introduced in this study. Recently, the production mode is shifting from mass customization to mass personalization (Qin and Lu 2021). Convertibility of manufacturing systems should hence be of more importance in order to meet the individual customization demand. Huang et al. (2019) concerned convertibility enhancement of RMS through delayed reconfiguration. Machine-level changeovers (Huang and Saitou 2020) provided a great option to realize convertibility rapidly and cost-effectively. Gadalla and Xue (2018) proposed an approach to identify optimal reconfiguration processes by considering the configuration variation of RMT. Touzout and Benyoucef (2019) proposed a multi-objective model for optimizing the reconfiguration process plan by considering the configuration changes of RMT, where an iterative multi-objective integer linear programming approach is developed to solve this model. Similarly, Khezri et al. (2021) attempted to generate reconfiguration process plan by minimizing sustainability-metric value, total production time, and total production cost using heuristic algorithms. Liu et al. (2019) proposed a distributed reconfiguration planning algorithm by considering modular robots that share the similar characteristics of RMT. Due to the narrower production time window and higher demand uncertainty, it is necessary and important to effectively handle dynamic arrival production tasks in a dynamic environment (Johnson et al. 2022). However, most of the existing studies focus on the offline method to figure out the reconfiguration planning problem of manufacturing systems resulting in static solutions generally. They fail to perceive the real-time status of SRMS and provide optimal solutions for reconfiguration planning dynamically. It is significant to update reconfiguration planning method and promote the production accuracy and efficiency of smart manufacturing systems with RMT at a high level. In addition, process planning (Khan et al. 2022), scheduling (Yang and Xu 2022), capacity planning (Alexopoulos et al. 2011), and other points should be considered when optimizing the performance of SMS. The single point optimization that largely exists in literature is gradually unable to meet the new demands for more and more complex SMS.

Digital twin, as a synthesis of emerging ICT and AI technologies, serves as a core enabler for SRMS. It enhances production efficiency and accuracy through bidirectional synchronization between virtual and physical spaces (Leng et al. 2021; Zhou et al. 2021; Tao et al. 2018). Notably, researchers emphasize that digital twin models must inherently possess reconfigurability to adapt to dynamic manufacturing needs (Zhang et al. 2021). To address this requirement, multiple systematic approaches have been proposed: First, Zhang et al. (2019) developed a reconfigurable strategy featuring an expandable model architecture with reserved interfaces for objective functions and optimization algorithms, enabling digital twin systems to satisfy multi-granularity reconfiguration targets. Leng et al. (2020) introduced a digital twin-driven rapid reconfiguration methodology through open architecture RMT implementations. In parallel, da Cunha et al. (2021) proposed a modular digital twin component design to facilitate reconfigurability in manufacturing systems, while Kombaya Touckia et al. (2022) established a dedicated simulation model framework for reconfigurable manufacturing systems (RMS). For optimization-oriented implementations, Ali et al. (2023) demonstrated how digital twins can resolve discrete multi-objective optimization problems in reconfigurable production lines. From a systemic perspective, Caesar et al. (2023) presented a holistic digital twin framework for reconfiguration management, addressing the interdependencies among system components. Complementing this, Haddou Benderbal et al. (2020) designed an integrated RMS digital twin (RMS-DT) modular framework that enables real-time system state representation and holistic performance optimization through flexible decision-making. Additionally, Yang et al. (2022) created a digital twin-based reconfiguration assessment platform that synchronizes with real-time manufacturing data, providing automated simulation capabilities for system analysis and reconstruction validation. Collectively, these advancements demonstrate that integrating reconfigurable digital twins with modular architectures and cross-layer optimization strategies not only enhances the adaptability of SMS, but also paves the way for achieving truly autonomous and resilient production ecosystems.

Traditionally, discrete optimization algorithms can solve production planning and reconfiguration problem well, such as Genetic Algorithm (Katoch et al. 2021), Simulated annealing (Kirkpatrick et al. 1983), Particle Swarm Optimization (Houssein et al. 2021), which is an offline mode and cannot meet the requirement of SMS. Leveraging its dynamic interaction with the environment during optimization, deep reinforcement learning offers a compelling solution for shifting reconfiguration planning from offline to online modes. This study actively explores effective reconfiguration planning methods to align with the rapid advancements in technology and adapt to emerging application scenarios. Reinforcement learning (Kaelbling et al. 1996; Lee and Choi 2020; Wang et al. 2019) is the problem faced by an agent that learns behavior through trial-and-error interactions with environment, which is learning what to do—how to map situations to actions—so as to maximize reward (Sutton and Barto 2018). Recently, deep learning has been prevailing in reinforcement learning in recent years to scale to decision-making problems with high-dimensional state and action spaces (Li 2017; Watkins and Dayan 1992; Arulkumaran et al. 2017). The application of reinforcement learning is becoming popular in manufacturing domain. Wang et al. (2019) adopted reinforcement learning to optimize the energy efficiency of Industrial Internet of Things (IIoT) that can promote management efficiency alongside with Industrial Internet platforms (Liu and Xie 2024). Yang and Xu (2022) used deep reinforcement learning to solve the scheduling and reconfiguration model in smart manufacturing. Epureanu et al. (2020) proposed a self-repair method based on deep reinforcement learning for SMS to find optimal strategy by considering system status and performance. Bakopoulos et al. (2024) studied the production scheduling based on deep reinforcement learning under the framework of digital twin. Therefore, deep reinforcement learning emerges as a powerful and adaptive approach, offering significant potential for addressing the dynamic reconfiguration and optimization challenges of SMS/SRMS.

In summary, the integration of digital twins and reinforcement learning provides an innovative solution for dynamic optimization and reconfiguration planning of intelligent manufacturing systems for reconfigurable machine tools. By constructing a high-fidelity digital twin model, real-time mapping and multi-dimensional data fusion of the physical manufacturing system can be realized, and equipment state, process flow and environmental parameters can be accurately captured, providing a virtual and real interactive simulation training environment for deep reinforcement learning algorithms. Under this framework, deep reinforcement learning can break through the limitations of traditional planning methods in dynamic disturbance response and complex constraint processing through independent exploration of strategy space, and significantly improve the efficiency and accuracy of system reconfiguration planning.

3 Problem Description

A typical SMS with the consideration of arch-type RMT based on digital twin is shown in Fig. 1a, where a couple of RMTs constitute a manufacturing system and digital twin technology is used to connect physical space and virtual space for enhancing the smart awareness of demand fluctuation and dynamically optimizing the performance of the manufacturing system. Due to the capability of digital twin, the smart manufacturing system can recognize the demand changes (as shown in red line in the left top of Fig. 1a) that can be the changes of part type, part quantity, or both in both physical space and virtual space, where reconfiguration planning is dynamically executed by considering the flexibility of production precedence and operation sequences simultaneously. Also, the reconfiguration activities will synchronously happen in physical and virtual space to update the performance of the manufacturing system.

There is more than one part to be produced in a smart manufacturing system. The production precedence of these parts could be changed due to some specific reasons (production precedence flexibility), which may require different manufacturing resources. Namely, the reconfiguration planning could be different when the production precedence of parts is changed. For example, as shown in Table 1, there are two parts (part a and part b) to be produced in a smart manufacturing system consisting of two RMTs (RMT 1 and RMT2). Part a should be produced by RMT 1 with configuration 2 (named c_{12}) and RMT 2 with configuration 1 (named c_{21}). Part b should be produced by c_{11} and c_{22}. Assuming the current configuration of these two RMTs are c_{12} and c_{21}, if part a is produced first, these two RMTs only need to be reconfigured when producing part b,$c_{12} \rightarrow c_{11}$ and $c_{21} \rightarrow c_{22}$. However, if part b is produced at first, reconfiguration activities will occur when processing part a and part b, $c_{12} \rightarrow c_{11} \rightarrow c_{12}$ and $c_{21} \rightarrow c_{22} \rightarrow c_{21}$. Therefore, it is significant

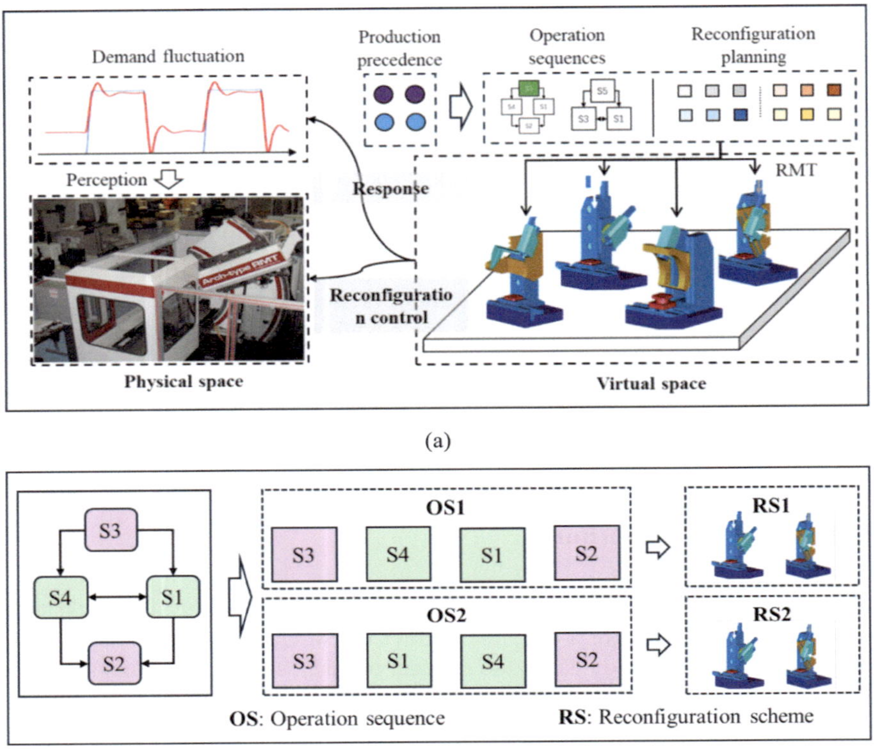

Fig. 1 Problem description for dynamic reconfiguration planning: **a** Digital twin-driven smart manufacturing system based on RMTs; **b** The influence of operation flexibility

to consider the flexibility of production precedence when executing reconfiguration planning.

Similarly, the operation sequence of a part is generally not rigid. In other words, the operation sequence of a part could be flexible. Different operation sequences may require different manufacturing resources resulting in different reconfiguration schemes as well. Consider the part with four features to be machined and the corresponding precedence order as shown in Fig. 1b. Due to whether S4 or S1 can be selected when S3 is completed, the example part can be machined using flexible

Table 1 The influence of production precedence flexibility

Production precedence	Smart manufacturing system	Reconfiguration planning
Part a → Part b	(c_{12}, c_{21})	$\rightarrow (c_{11}, c_{22})$
Part b → Part a		$\rightarrow (c_{11}, c_{22}) \rightarrow (c_{12}, c_{21})$

operation sequences (S3-S4-S1-S2 or S3-S1-S4-S2) that can lead to different reconfiguration schemes. Moreover, there is more than one option when selecting RMT for producing a specific feature of a part. That is to say, the same feature of a part can be produced in different RMTs producing different reconfiguration planning results. Take the part in Fig. 1b as an example again, although there is no precedence limitation between S4 and S1, the reconfiguration options (different RMTs or configurations of the same RMT) could affect operation sequences as well. If the configuration for S3 can be used for machining S1, operation sequence 2 could be the better one. Thus, it is also important to exploit the operation flexibility of parts when conducting reconfiguration planning for SRMS.

The reconfiguration planning problem of SRMS by considering the flexibility of production precedence and operation sequences simultaneously can be described as follows. A smart manufacturing system with RMT is constructed to possess the ability of handling dynamic demands, where $SMS = \{RMT_m | m = 1, 2, 3, \ldots\}$ denotes the structure of the smart manufacturing system and $PT = \{(P_i, N_i) | i = 1,2, \ldots\}$ means the dynamic production tasks assigned to the smart manufacturing system by providing the information of part type P_i and part number N_i. Part i consists of several features to be produced, that is, $f_j^i (j = 1, 2, \ldots)$, where priority relationships between these features is given and will be considered during optimization. When production task PT changes, the smart manufacturing system should simultaneously find the optimal reconfiguration scheme, the production precedence and operation sequences, which the reconfiguration effort of RMTs themselves, part moving effort among RMTs, and processing cost are used to estimate the optimization process. So, the fundamental problem lies in dynamically reconfiguring the RMTs in SRMS efficiently and accurately, while simultaneously optimizing production precedence and operation sequences.

4 Reconfiguration Planning Method Based on Reinforcement Learning

4.1 *Assumptions for Reconfiguration Planning Issue*

The following necessary assumptions are made: A machine can process only one part at a specific time, and only one feature of a part can be processed at a time. A Reconfigurable Machine Tool (RMT) can be reconfigured into multiple configurations, and each feature can be completed by a specific configuration of the RMT. Once the processing of a feature begins in a particular RMT configuration, it will continue uninterrupted until completion. A part, composed of multiple features, is considered complete once all its features have been processed. Additionally, each RMT can process only one part at a time. To control the complexity of the problem, mixed-model production is not considered in this study.

4.2 Reconfiguration Process Modeling for SRMS

The reconfiguration process of SRMS is a typical MDP, which is executed based on the current state without necessarily considering the influence of the previous states. And the interaction between a learning agent and its environment in terms of states, actions, and rewards should be defined for reinforcement learning using MDP. Reinforcement learning typically employs MDP to formalize the interaction between an agent and its environment through states, actions, and rewards. The reconfiguration process modeling will sort out the details of the MDP tuple $< S, A, P, R, \gamma >$, where S denotes state space, A represents action space, P denotes state transition probability, R represents the reward function, and $\gamma \in (0,1]$ is reward discount factor. And the state transition probability $P(s\prime|s, a) = 1$, because the next state is always fixed when action is taken in the current state in the work.

(1) State space of SRMS

In SRMS, parts with one or more features will be assigned to one or more RMTs and different configurations could be needed. At time slot t, the system state consists of (1) part processing progress, (2) feature processing progress, (3) part position, and (4) configuration of RMT. So, the current state $s_t \in S$ can be expressed by Eq. (1). Additionally, the agent will take appropriate action to update the current state during production.

$$s_t = \boldsymbol{U}_t, \boldsymbol{F}_t, \boldsymbol{Pos}_t, \boldsymbol{C}_t \tag{1}$$

where, $\boldsymbol{U}_t = \{u_i|i = 1, 2, \ldots\}$ denotes the part processing progress at time t, u_i records the processing progress of part i ($u_i \in \{0, 1, 2\}$ represents *NotStarted*, *InComplete* and *Completed* states respectively of part i).

For example, $\boldsymbol{U}_t = \{0, 1, 2, 2\}$ means the production of part 1 has not started, part 2 is incomplete, while part 3 and part 4 have been completed at time t. $\boldsymbol{F}_t = \left\{f_j^i|i = 1, 2, \ldots; j = 1, 2, 3\ldots\right\}$ represents the processing progress of feature j in part i at time t ($f_j^i = 1$ means feature j of part i is completed, otherwise$f_j^i = 0$).

For example, part 2 is incomplete at time t, the corresponding $\boldsymbol{F}_t = \{1, 0, 0, 1\}$ means feature 1 and feature 4 has been completed, while feature 2 and feature 3 are incomplete at time t. $\boldsymbol{Pos}_t = \{p_m|m = 1, 2, \ldots\}$ is used to show the position of part i at time t, where p_m means the fixed position of RMT m in the smart manufacturing system.

Due to only the position changes of parts being concerned, the real position value will not affect the results. Namely, p_1 means the position of the RMT1, which the real position value RMT1 will not involve. As mentioned before, part 2 is incomplete, if $\boldsymbol{Pos}_t = \{p_2\}$ means part 2 is located in RMT2. $\boldsymbol{C}_t = \{c_{mk}|m = 1, 2, \ldots; k = 1, 2, \ldots\}$ is a configuration set of the RMTs in the current manufacturing system at time t, where c_{mk} refers to configuration k of RMT m. For example, $\boldsymbol{C}_t = \{c_{11}, c_{21}, c_{32}\}$ means there are three RMTs in the smart manufacturing system and the configuration state of RMT1, RMT2, and RMT3 are configuration 1 (c_{11}), configuration 1 (c_{21}), and configuration 2 (c_{32}) respectively.

(2) Action space of SRMS

In reconfiguration process, the action can be described from two aspects: (1) Selecting a new part, namely wp^i; (2) Selecting a feature of the current part (*InComplete*) to process using a specific configuration of the selected RMT, namely $c_{mk}^{f_j^i}$, where $c_{mk}^{f_j^i}$ means the agent select RMT m with configuration k for processing feature j of part i. Reconfiguration activities could happen if the current configuration of the selected RMT is not the selected one. Based on the description of the state and action, the decision point at which the agent selects an action to update the state should be after a feature of the current part is completed or after a new part is selected. So, the action $a_t \in A$ can be defined according to Eq. (2).

$$a_t = \begin{cases} wp^i, \ Condition\ 1 \\ c_{mk}^{f_j^i}, \ Condition\ 2 \end{cases} \tag{2}$$

where *Condition* 1 means when current part is completed or the system is in initial state, the agent will select a new part for production from the current production task. *Condition* 2 means a part has been selected in state (*InComplete*), action $c_m k^{\left(f_j^i\right)}$ will be taking to process a new feature of the selected part $\left(f_j^i\right)$, which can be completed by usingc_{mk}. For example, a part contains two features and a feature can be processed in different RMTs, the mapping among features and configurations is shown in Fig. 2a, where the corresponding action space is $\left\{wp^1, c_{11}^{f_1^1}, c_{21}^{f_1^1}, c_{12}^{f_2^1}\right\}$. Similarly, a part contains two features as shown in Fig. 2b, where the corresponding action space is $\left\{wp^2, c_{11}^{f_1^2}, c_{21}^{f_2^2}, c_{22}^{f_2^2}, c_{31}^{f_3^2}\right\}$.

The process of reconfiguration from the current state to the next state by conducting the selected action is regarded as a policy $\pi : S \rightarrow A$, which is the

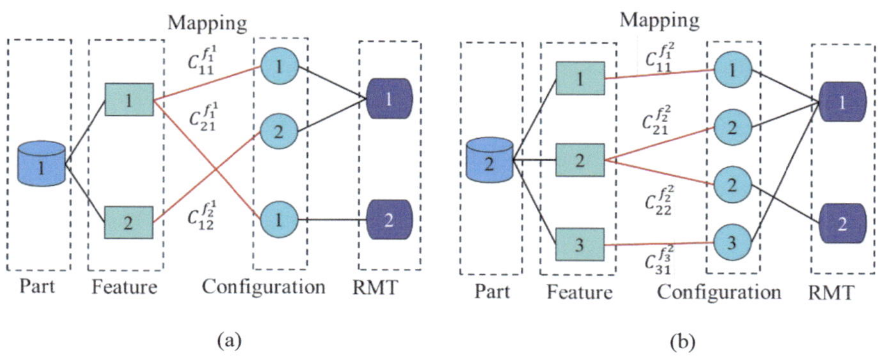

Fig. 2 Action space example: **a** Example; **b** Example 2

mapping between the state space and the action space. The policy provides a definition of action selection of agent in current state. When finishing performing action a_t in state s_t, the next state $s_{t+1} \Leftarrow s_t + a_t$ can be obtained. Adopting the example in Fig. 2 again, supposing the current state is $s_1 = \{u_1 = 0\}, \{f_1^1 = 1, f_2^1 = 0\}, p_1, \{c_{11}, c_{21}\}$, the agent will choose the action a_1 from action space according to the policy π. In this situation, $a_1 = c_{22}^{f_2^1}$ will be the best action.

(3) Reward modeling for SRMS

Generally, a reward will be obtained when agent executes an action under a state, that is, $r_t \sim R(s_t, a_t)$. The reward determines the optimization direction of deep reinforcement learning, which is the crucial indicator for evaluating the optimization effectiveness. Therefore, the reward function for reconfiguration planning will be set based on the minimum production cost of multiple parts, as shown in Eq. (3).

$$R(s_t, a_t) = \begin{cases} GP - \begin{pmatrix} \delta \cdot d(p_m, p_{m'}) \\ +g(c_{m'k}, c_{m'k'}) \\ +N_i \cdot pd\left(c_{m'k}^{f_j^i}\right) \end{pmatrix}, & (s_t, a_t) \in Condition\ 3 \\ positive\ value, & (s_t, a_t) \in Condition\ 4 \\ negative\ value, & (s_t, a_t) \in Condition\ 5 \end{cases} \tag{3}$$

where *Condition* 3 denotes selecting a feature of the current part for producing, which the production cost will be calculated. *GP* denotes the gross profit of producing a specific production task. $d(p_m, p_{m'})$ means the cost of moving part from RMT_m to $RMT_{m'}$ (if the part is processed in the same RMT $((\text{i.e.,}\ m = m'),\ d(p_m, p_{m'}) = 0)$. $\delta = \lfloor N_i/B \rfloor$ is moving factor considering parts can be moved in batch for saving cost, which N_i is total number of part$i$, B denotes batch size, and $\lfloor\ \rfloor$ means round up to an integer. $g(c_{m'k}, c_{m'k'})$ means the reconfiguration cost of $RMT_{m'}$ from configuration k to configuration k' (similarly, if the configuration of $RMT_{m'}$ is exactly needed (reconfiguration is not needed), $g(c_{m'k}, c_{m'k'}) = 0)$. $pd\left(c_{m'k}^{f_j^i}\right)$ means processing cost of producing feature i of part j in configuration k of $RMT_{m'}$. *Condition* 4 is selecting an unprocessed part from the production task when initializing production or the current part is completed, which a *positive value* (the value should greater than zero) will be assigned to reward this action. *Condition* 5 means other situations, including selecting the same part, reprocessing the same feature, violating operation sequence, etc., in which a negative value (the value should less than zero) will be assigned to penalize this action. Above all, the reward is obtained based on the evaluation of actions, which can be explained through two phases: (1) An action will meet one of the three different conditions. (2) The action will be evaluated under a specific condition. In *Condition* 3, the benefit of completing a production task will be calculated by considering the gross profit, processing cost, reconfiguration cost, and moving cost. In *Condition* 4, if a right part is selected, a positive value will be assigned *Condition* 5 means a wrong action will be punished.

4.3 Reconfiguration Planning Based on DQN

As a classical reinforcement learning algorithm, Q-learning (Kaelbling et al. 1996) optimizes policies by constructing a Q-table to store expected returns for state-action pairs. However, its tabular approach requires exhaustive storage of Q-values for all possible state-action combinations, leading to curse of dimensionality. The Deep Q-Network (DQN) (Lee and Choi 2020; Wang et al. 2019) addresses this limitation by replacing discrete Q-tables with continuous parametric Q-functions through deep neural network approximators. This methodology leverages the nonlinear mapping capabilities and automated feature extraction of neural networks to learn state representations directly from raw input data (e.g., image pixels), thereby enabling effective generalization across large-scale continuous state spaces. Hence, DQN is adopted to solve the proposed reconfiguration process model in this work. The optimization details of DQN for reconfiguration planning are shown in Fig. 3.

DQN enables agents to maximize cumulative rewards during environmental interactions. Q-value function represents the long-term expected return of an agent performing an action in a given state, as shown in Eq. (4).

$$\begin{aligned} Q^{\pi}(s,a) &= E_{\pi}[G_t|s_t = s, a_t = a] = E_{\pi}\left[r_{t+1} + \gamma\left(r_{t+2} + \cdots + \gamma^{k-1} r_{t+k}\right)|s_t, a_t\right] \\ &= E_{\pi}\left[r_{t+1} + \gamma Q^{\pi}(s_{t+1}, a_{t+1})|s_t, a_t\right] \end{aligned} \tag{4}$$

where return G_t is the total discounted reward of the agent from time t to the end. γ is the discount factor which used to trade off the influence of future reward on return. It is also capable of ensuring the convergence of G_t.

To guarantee the efficiency of training, the Temporal-Difference method based on the Bellman expectation equation is adopted to execute single-step update, as shown in Eq. (5).

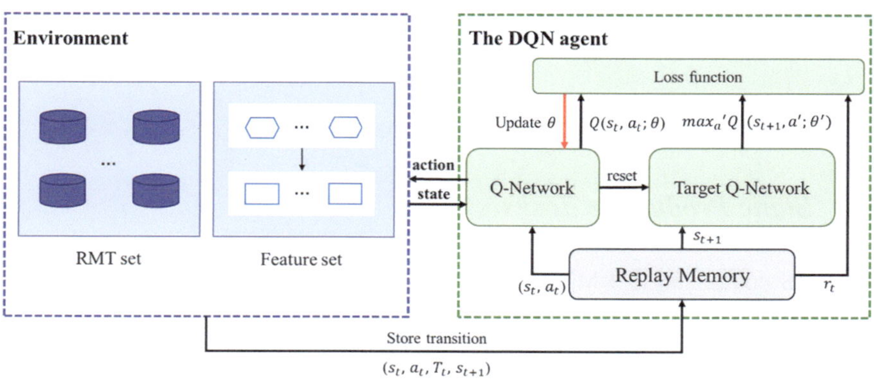

Fig. 3 Reconfiguration planning based on DQN

$$Q^{new}(s_t, a_t) \leftarrow (1-\alpha)Q^{old}(s_t, a_t) + \alpha\left[r_{t+1} + \gamma \max_{a' \in A} Q^{\pi}\left(s_{t+1}, a'\right)\right] \tag{5}$$

where s_{t+1} and a' are the next state and action, respectively. $\alpha \in (0,1)$ is the learning rate, which determines the influence of new information on current Q-value. A lower learning rate may require more learning rounds due to smaller influence, however, a higher learning rate could lead to sub-optimal and even cannot converge.

This formulation enables efficient policy derivation by selecting actions that maximize the predicted Q-values, while maintaining tractability in high-dimensional spaces through neural networks' inherent capacity for non-linear function approximation and hierarchical feature extraction. The agent subsequently interacts with the environment through an exploration–exploitation strategy, collecting transition experiences that are utilized for iterative network optimization via gradient-based minimization of temporal difference errors, thereby progressively aligning Q-value predictions with the Bellman optimality criterion. These experiences are used to update the parameters $\boldsymbol{\theta}$ of Q-Network to make it more accurate as shown in Eq. (6)

$$\boldsymbol{\theta} \leftarrow \boldsymbol{\theta} - \eta \nabla_{\theta} L_{\theta} \tag{6}$$

where η is the learning rate that controls the weight updating speed of the neural network; ∇ denotes the gradient function, and L is the loss function as shown in Eq. (7)

$$L_{\theta} = \frac{1}{2}[target - prediction]^2 = \frac{1}{2}\left[r_{t+1} + \gamma \max_{a' \in A} Q_{\theta}\left(s_{t+1}, a'\right) - Q_{\theta}(s_t, a_t)\right]^2 \tag{7}$$

Besides, the techniques of experience replay and target network are used to help DQN agent stabilize the training process and solve the problem of sample correlation and instability in reinforcement learning. The pseudocode of DQN for reconfiguration planning is given in Table 2.

5 Case study for Various Production Tasks

5.1 *Static Production Tasks*

A SRMS comprising 5 RMTs is utilized to examine the reconfiguration process for static production tasks, which involving four parts (Production Task 1) and is unchanged. The corresponding digital twin of SRMS is developed on the Unity 3D platform to create a high-fidelity virtual environment and enable real-time monitoring of production activities, as illustrated in Fig. 4. Additionally, the feature precedence of the four parts and the configuration-feature mapping are provided in Fig. 5.

Table 2 DQN pseudocode

Inputs: Reconfiguration planning problem

Output: Parameter $\boldsymbol{\theta}$ for Q-Network

Initialize replay memory D to capacity N

Initialize Q-Network with weight $\boldsymbol{\theta}$

Initialize target Q-Network with weights $\boldsymbol{\theta}'$

For episode = 1, $\boldsymbol{M}$ **do**:

    Initialize state $s = s_0$;

    For t=1, $\boldsymbol{T}$ **do**

        With probability $\boldsymbol{\varepsilon}$ select a random action a_t;

        Otherwise select $a_t = argmax_{a \epsilon A} Q^*(s_t, a)$;

        Execute action a_t in emulator

        Observe reward r_t and new state s_{t+1};

        Save transition (s_t, a_t, r_t, s_{t+1}) in D

        Sample random batch of transitions (s_j, a_j, r_j, s_{j+1}) from D

        Set $y_j = \begin{cases} r_j, & for\ terminal\ s_{j+1} \\ r_j + \gamma max_{a'} Q(s_{j+1}, a'; \boldsymbol{\theta}), & for\ non-terminal\ s_{j+1} \end{cases}$

        Perform a gradient decent regarding weights $\boldsymbol{\theta}$

        Every C steps reset $\theta' = \theta$

    End for

End for

The optimization goal of this case is to maximize the return of completing the production task in the selected smart manufacturing system with 5 RMTs. *GP* value and the batch size of production can be set according to the feature of production task. The necessary parameters should be preset before optimization, that is, *positive value* = 1, *negative value* = −50. In addition, the moving cost and the reconfiguration cost are given in Fig. 6. The red arrow means reconfiguration direction. The number on the red arrow means the value of reconfiguration cost. The processing cost is given in Table 3.

The optimal solution based on three randomly initial situations of the smart manufacturing system with 5 RMTs is explored to inspect the effectiveness and adaptability of the proposed method. The corresponding configurations of these 5 RMTs are given in Table 4. Additionally, the parameters of DQN must be carefully configured for the three situations, drawing on insights from existing literature and the specific details of each situation, as outlined in Table 5. To explore more states and actions at the beginning of training, the initial epsilon value is set as $\varepsilon = 0.9$, which gradually decrease linearly to 0.1 for increasing the utilization of the learned strategies.

The training processes for these experiments are illustrated in Fig. 6. In DQN, the loss value of the neural network serves as a metric to quantify the discrepancy between the predicted value and the target value, indicating the prediction accuracy of the Q-value for the current state. Additionally, the value function is deemed to have converged when the change in the return function becomes sufficiently small. Also,

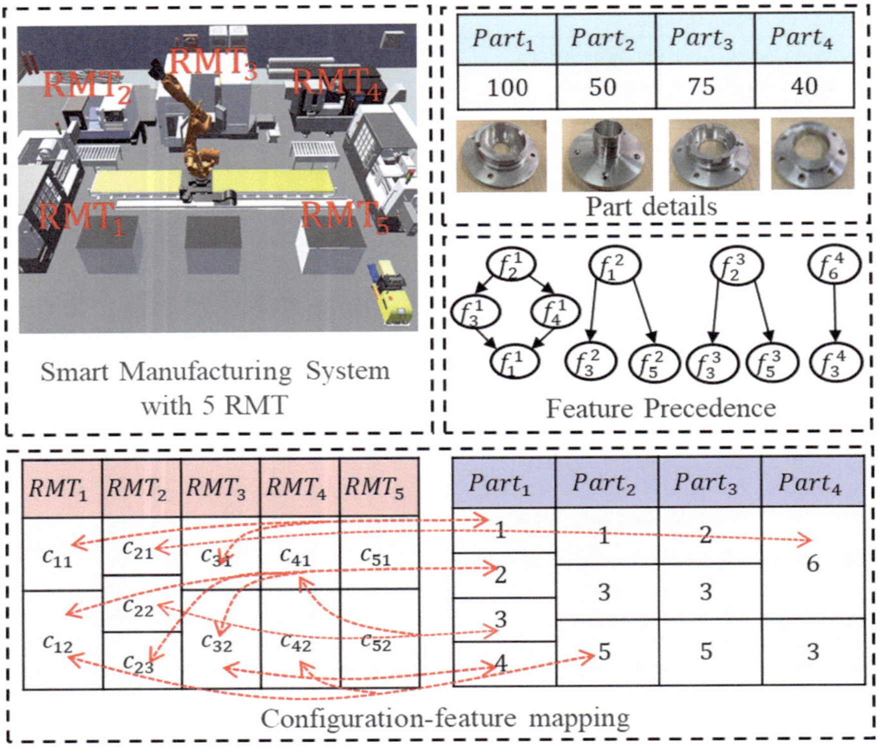

Fig. 4 Case study scenario and production task 1

if the change in the return function is small, the value function is considered to have converged. Therefore, the training effectiveness and convergence of DQN can be determined by evaluating the changes in the return value curve and loss value curve. The ε value (Light green curve) in all experiments decreases gradually to sufficiently explore the potential solutions until ε value equals 0.1. The return value (Light orange curve) and loss value (Bule curve) in all experiments can converge a stable level within 400 rounds, indicating that the agent's strategy or value function has converged. In additionally, The experiments for these three situations are conducted on PyCharm with Python, running on a 3.20 GHz AMD Ryzen 7-5800H laptop.

The experiment results are shown in Table 6, including returns and optimal policies. The returns of these three situations are 713.4, 718.4, and 706.2 respectively. The policy of these experiments includes the description of the production sequence of parts, the operation sequence of each part and the reconfiguration sequence of RMTs. As shown in Fig. 7, in Situation 1, the production sequence of parts is *part* 2 $\rightarrow$ *part* 3 $\rightarrow$ *part* 1 $\rightarrow$ *part* 4; The operation sequence of *part* 2 is to process *feature* 1 $\rightarrow$ *feature* 3 $\rightarrow$ *feature* 5 using RMTs c_{11}, c_{22} and c_{52} in turn, and *part* 3, *part* 4 and *part* 1 follow the same pattern; The reconfiguration activity

Moving Cost					
	RMT_1	RMT_2	RMT_3	RMT_4	RMT_5
RMT_1	0	2	1.2	3.8	4
RMT_2	2	0	2.1	3	1
RMT_3	1.2	2.1	0	0.8	1.9
RMT_4	3.8	3	0.8	0	2.3
RMT_5	4	1	1.9	2.3	0

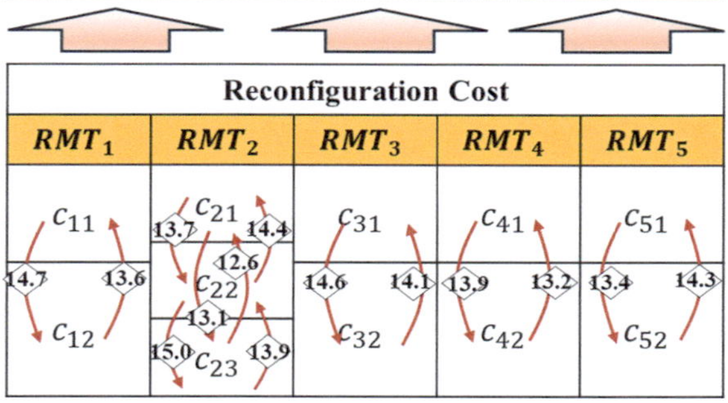

Fig. 5 Moving cost and reconfiguration cost

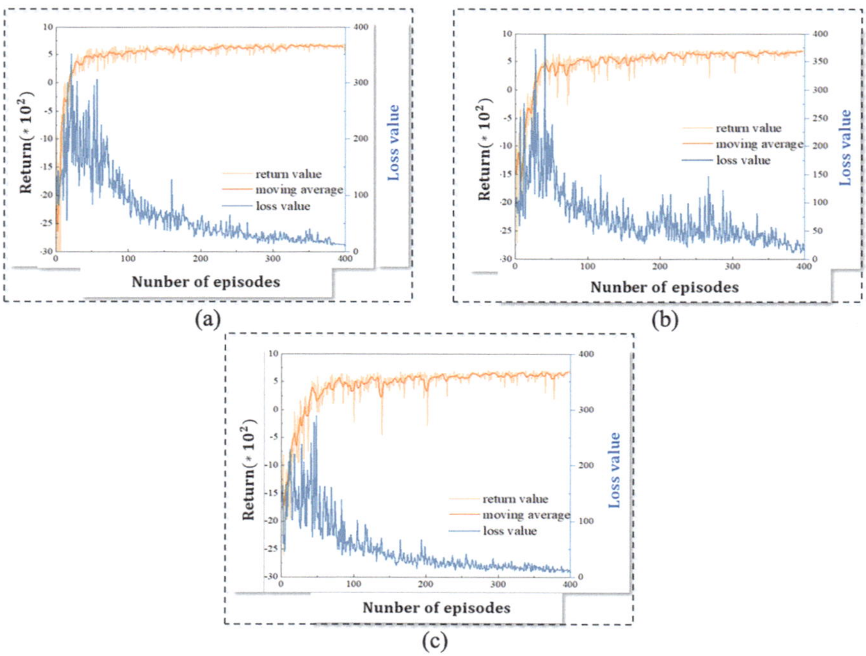

Fig. 6 Training process of three situations: **a** Situation 1; **b** Situation 2; **c** Situation 3

Table 3 Processing cost

Parameter	Value	Parameter	Value	Parameter	Value
$pd\left(c_{11}^{f_1^{1(2)}}\right)$	0.68(0.59)	$pd\left(c_{22}^{f_3^{1(2)(3)(4)}}\right)$	0.54(0.51) (0.47)(0.44)	$pd\left(c_{12}^{f_5^{2(3)}}\right)$	0.74(0.47)
$pd\left(c_{31}^{f_1^{1(2)}}\right)$	0.63(0.68)	$pd\left(c_{41}^{f_3^{1(2)(3)(4)}}\right)$	0.60(0.33) (0.65)(0.40)	$pd\left(c_{42}^{f_5^{2(3)}}\right)$	0.42(0.59)
				$pd\left(c_{52}^{f_5^{2(3)}}\right)$	0.30(0.55)
$pd\left(c_{12}^{f_2^{1(3)}}\right)$	0.33(0.43)	$pd\left(c_{32}^{f_4^{1}}\right)$	0.32		
$pd\left(c_{23}^{f_2^{1(3)}}\right)$	0.75(0.37)	$pd\left(c_{51}^{f_4^{1}}\right)$	0.59	$pd\left(c_{21}^{f_6^{4}}\right)$	0.50
$pd\left(c_{32}^{f_2^{1(3)}}\right)$	0.37(0.55)				

Table 4 Randomly initial configurations of 5 RMTs

RMT	Situation 1	Situation 2	Situation 3
RMT_1	c_{11}	c_{12}	c_{11}
RMT_2	c_{22}	c_{21}	c_{21}
RMT_3	c_{32}	c_{31}	c_{31}
RMT_4	c_{41}	c_{42}	c_{41}
RMT_5	c_{52}	c_{52}	c_{51}

Table 5 DQN parameter settings of three situations

Parameter	Value
Learning rate η	0.01
Discount factor γ	0.95
Epsilon ε	0.9
EPISODES M	400
Replay buffer size N	3000
Batch size B	128
Update frequency C	100

is $c_{22} \rightarrow c_{21}$, means that RMT2 is reconfigured from c_{22} to c_{11} to process feature 6 of *part* 4. Similarly, the optimal policy of Situation 2 and Situation 3 can be obtained from Fig. 7 as well. It can be observed that these reconfiguration schemes never violate feature precedence in Fig. 4 and only require reconfiguration activities when necessary. Besides, the reconfiguration planning results demonstrate that the proposed method can fully facilitate the initial configurations to reduce reconfiguration activities to avoid frequent production interruption, which empowers the smart manufacturing system with RMT to deal with demand fluctuation easily.

5.2 Dynamic Production Tasks

To further demonstrate the application process of the proposed method and validate its effectiveness, a series of experiments involving dynamic production tasks are conducted. The specifics of these dynamic production tasks are detailed in Fig. 8. The part quantities in the new production tasks vary, where a zero quantity indicates that the part is not required for the task. These four new production tasks are assumed to arrive sequentially, starting from the final states of Situation 1 in Production Task 1, as shown in Fig. 4. This setup also allows for testing the robustness of the trained DQN model. The optimal policy of these new production tasks is shown on Fig. 8. These results based on the trained DQN model for different production tasks do not violate the assumptions and restrictions in this work. That means the trained DQN model can deal with different production tasks well.

6 Conclusion

A online optimization method of reconfiguration planning for SRMS based on DQN is proposed in this chapter to dynamically adapt and optimize production activities when demand changes. And the main conclusion of this chapter are as follows:

(1) A digital twin for SRMS enables online reconfiguration planning optimization through timely demand change recognition, facilitating dynamic reconfiguration.
(2) The reconfiguration processes of SRMS are directly modeled based on MDP to construct an online optimization environment, where the flexibility of production precedence and operation sequences is included to promote the accuracy of the reconfiguration scheme.
(3) A deep reinforcement learning algorithm named DQN is adopted to dynamically search the optimization reconfiguration scheme of smart manufacturing systems with RMT, which the capability of DQN to efficiently solve high-dimensional problems can equip the proposed optimization model with high adaptability for handing dynamic demand fluctuation.

Table 6 Experiment results of three situations

Situation	Return	Optimal policy π^*
Situation 1	717.4	part 2 $\left[c_{11}^{f_1^2} \rightarrow c_{22}^{f_3^2} \rightarrow c_{52}^{f_5^2}\right] \rightarrow$ part 3 $\left[c_{32}^{f_2^3} \rightarrow c_{22}^{f_3^3} \rightarrow c_{52}^{f_5^3}\right] \rightarrow$ part 1 $\left[c_{22}^{f_3^1} \rightarrow c_{32}^{f_4^1} \rightarrow c_{11}^{f_1^1} \rightarrow c_{32}^{f_2^1}\right] \rightarrow$ part 4 $\left[c_{21}^{f_6^4} \rightarrow c_{41}^{f_3^4}\right]$
Situation 2	718.4	part 4 $\left[c_{21}^{f_6^4} \rightarrow c_{22}^{f_3^4}\right] \rightarrow$ part 2 $\left[c_{31}^{f_1^2} \rightarrow c_{52}^{f_5^2} \rightarrow c_{41}^{f_3^2}\right] \rightarrow$ part 3 $\left[c_{12}^{f_2^3} \rightarrow c_{12}^{f_5^3} \rightarrow c_{22}^{f_3^3}\right] \rightarrow$ part 1 $\left[c_{22}^{f_3^1} \rightarrow c_{31}^{f_1^1} \rightarrow c_{32}^{f_4^1} \rightarrow c_{12}^{f_2^1}\right]$
Situation 3	706.2	part 4 $\left[c_{21}^{f_6^4} \rightarrow c_{41}^{f_3^4}\right] \rightarrow$ part 2 $\left[c_{11}^{f_1^2} \rightarrow c_{41}^{f_3^2} \rightarrow c_{52}^{f_5^2}\right] \rightarrow$ part 3 $\left[c_{23}^{f_2^3} \rightarrow c_{22}^{f_3^3} \rightarrow c_{52}^{f_5^3}\right] \rightarrow$ part 1 $\left[c_{22}^{f_3^1} \rightarrow c_{31}^{f_1^1} \rightarrow c_{32}^{f_4^1} \rightarrow c_{32}^{f_2^1}\right]$

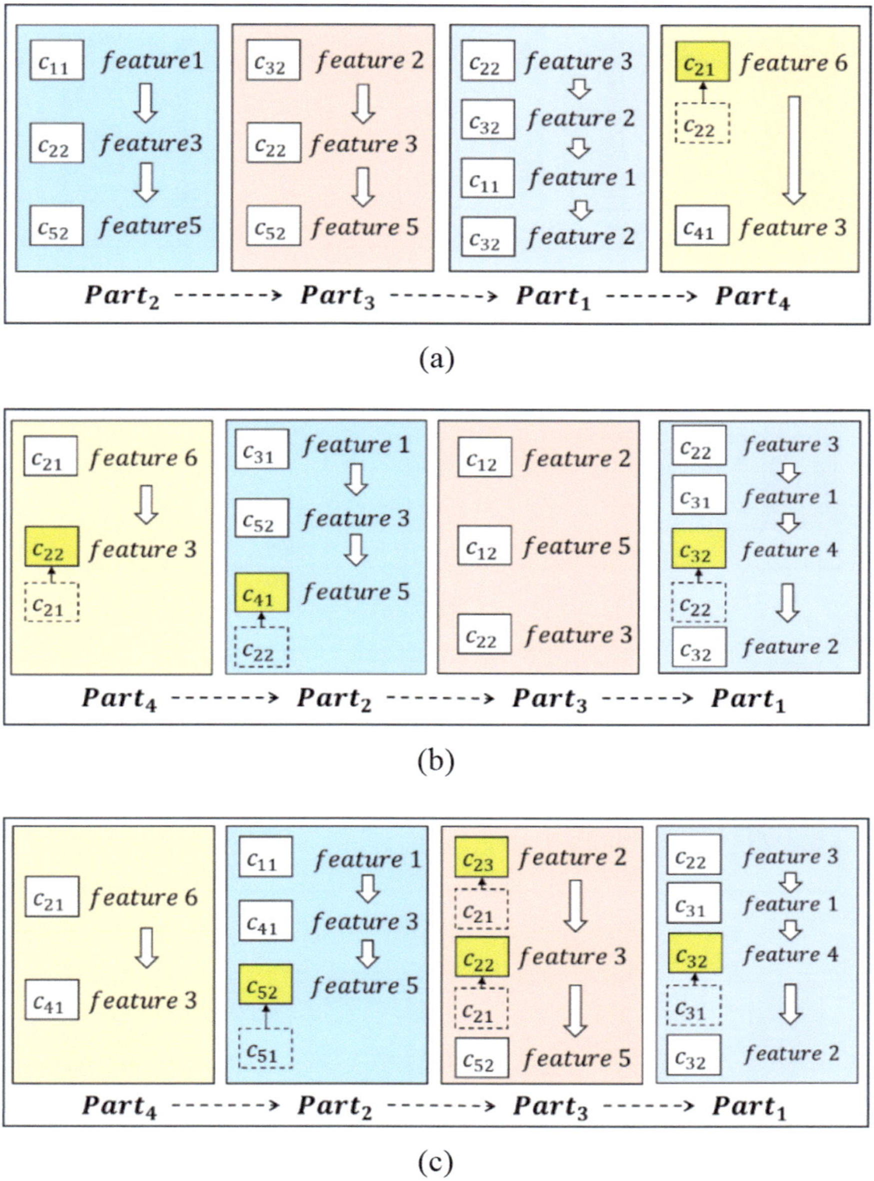

Fig. 7 The optimal policy of these three situations: **a** Situation 1; **b** Situation 1; **c** Situation 1

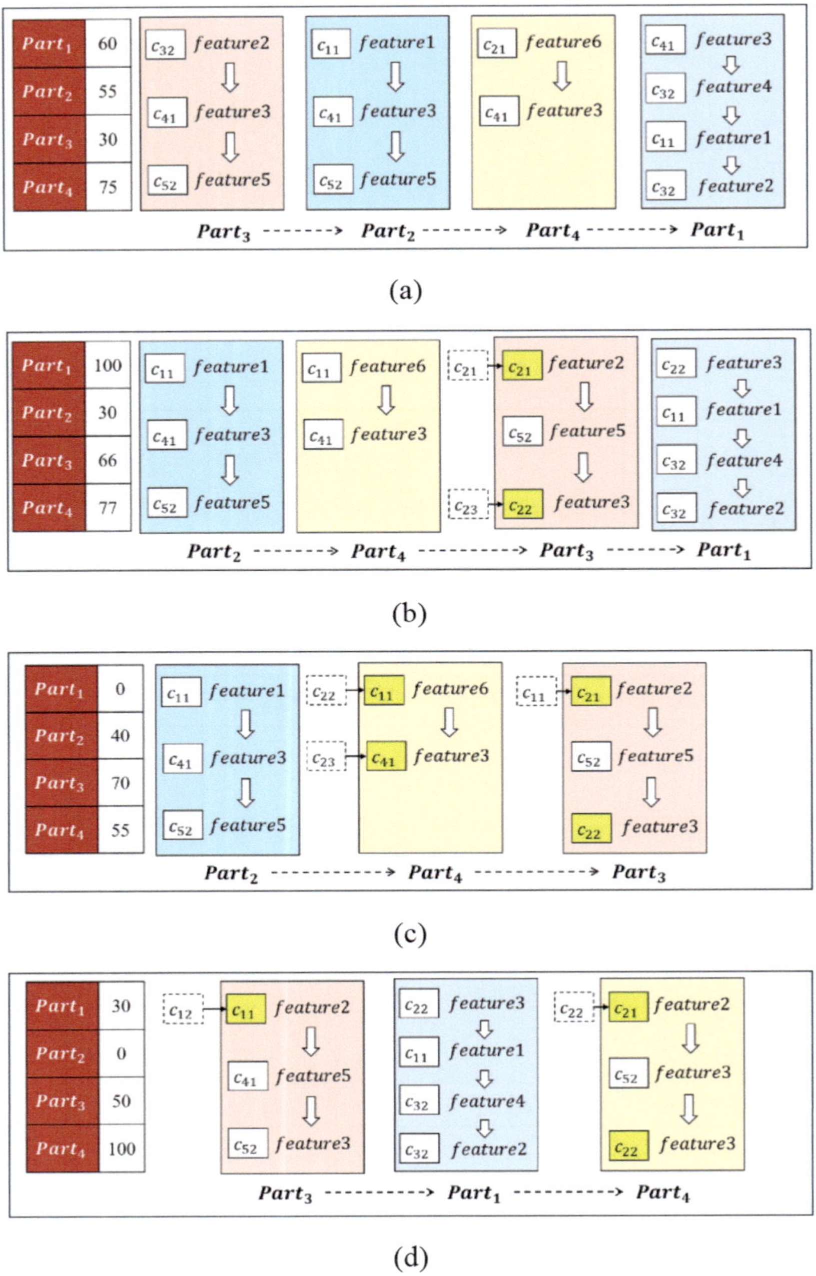

Fig. 8 The optimal policy of new production tasks: **a** Production Task 2; **b** Production Task 3; **c** Production Task 4; **d** Production Task 5

References

Ahmed I, Jeon G, Piccialli F (2022) From artificial intelligence to explainable artificial intelligence in Industry 4.0: a survey on what, how, and where. IEEE Trans Ind Inform 18(8):6069–6078

Alexopoulos K, Papakostas N, Mourtzis D et al (2011) A method for comparing flexibility performance for the lifecycle of manufacturing systems under capacity planning constraints. Int J Prod Res 49(11):3307–3317

Ali MA, Alarjani A, Mumtaz MA (2023) A NSGA-II based approach for multi-objective optimization of a reconfigurable manufacturing transfer line supported by digital twin: a case study. Adv Prod Eng & Manag 18(1):116–129

Arulkumaran K, Deisenroth MP, Brundage M et al (2017) Deep reinforcement learning: a brief survey. IEEE Signal Process Mag 34(6):26–38

Azab A, ElMaraghy HA (2007) Mathematical modeling for reconfigurable process planning. CIRP Ann 56(1):467–472

Bakopoulos E, Siatras V, Mavrothalassitis P et al (2024) Digital-twin-enabled framework for training and deploying AI agents for production scheduling. In: Artificial intelligence in manufacturing: enabling intelligent, flexible and cost-effective production through AI, pp 147–179. Springer Nature Switzerland, Cham

Bortolini M, Galizia FG, Mora C (2018) Reconfigurable manufacturing systems: literature review and research trend. J Manuf Syst 49:93–106

Caesar B, Barton K, Tilbury DM, Fay A (2023) Digital twin framework for reconfiguration management: concept & evaluation. IEEE Access 11:127364–127387

da Cunha C, Cardin O, Gallot G, Viaud J (2021) Olivier cardin, guillaume gallot, julien viaud, designing the digital twins of reconfigurable manufacturing systems: application on a smart factory. IFAC-PapersOnLine 54(1):874–879

Dhupia J, Powalka B, Katz R et al (2007) Dynamics of the arch-type reconfigurable machine tool. Int J Mach Tools Manuf 47(2):326–334

Epureanu BI, Li X, Nassehi A et al (2020) Self-repair of smart manufacturing systems by deep reinforcement learning. CIRP Ann 69(1):421–424

Gadalla M, Xue D (2018) An approach to identify the optimal configurations and reconfiguration pro cesses for design of reconfigurable machine tools. Int J Prod Res 56(11):3880–3900

Haddou Benderbal H, Yelles-Chaouche AR, Dolgui A (2020) A digital twin modular framework for reconfigurable manufacturing systems. In: Lalic B, Majstorovic V, Marjanovic U, von Cieminski G, Romero D (eds) Advances in production management systems. Towards smart and digital manufacturing. APMS 2020, p 592. IFIP Advances in Information and Communication Technology

Houssein EH, Gad AG, Hussain K et al (2021) Major advances in particle swarm optimization: theory, analysis, and application. Swarm Evol Comput 63:100868

Huang S, Wang G, Yan Y (2019) Delayed reconfigurable manufacturing system. Int J Prod Res 57(8):2372–2391

Huang S, Wang G, Yan Y (2022) Building blocks for digital twin of reconfigurable machine tools from design perspective. Int J Prod Res 60(3):942–956

Huang S, Saitou K (2020) Configuration design for make-to-order production considering individual or der arrivals and machine level changeovers. In: 2020 IEEE 16th international conference on automation science and engineering (CASE), pp 630–635. IEEE

Johnson D, Chen G, Lu Y (2022) Multi-agent reinforcement learning for real-time dynamic production scheduling in a robot assembly cell. IEEE Robot Autom Lett 7(3):7684–7691

Kaelbling LP, Littman ML, Moore AW (1996) Reinforcement learning: a survey. J Artif Intell Res 4:237–285

Katoch S, Chauhan SS, Kumar V (2021) A review on genetic algorithm: past, present, and future. Multimed Tools Appl 80:8091–8126

Khan AS, Homri L, Dantan JY et al (2022) An analysis of the theoretical and implementation aspects of process planning in a reconfigurable manufacturing system. Int J Adv Manuf Technol 119(9–10):5615–5646

Khezri A, Benderbal HH, Benyoucef L (2021) Towards a sustainable reconfigurable manufacturing system (SRMS): multi-objective based approaches for process plan generation problem. Int J Prod Res 59(15):4533–4558

Kirkpatrick S, Gelatt CD Jr, Vecchi MP (1983) Optimization by simulated annealing. Science 220(4598):671–680

Kombaya Touckia J, Hamani N, Kermad L (2022) Digital twin framework for reconfigurable manufacturing systems (RMSs): design and simulation. Int J Adv Manuf Technol 120:5431–5450

Koren Y (2013) The rapid responsiveness of RMS. Int J Prod Res 51(23–24):6817–6827

Koren Y, Heisel U, Jovane F et al (1999) Reconfigurable manufacturing systems. CIRP Ann 48(2):527–540

Koren Y, Wang W, Gu X (2017) Value creation through design for scalability of reconfigurable manufacturing systems. Int J Prod Res 55(5):1227–1242

Koren Y, Gu X, Guo W (2018) Reconfigurable manufacturing systems: principles, design, and future trends. Front Mech Eng 13:121–136

Kusiak A (2018) Smart manufacturing. Int J Prod Res 56(1–2):508–517

Lee S, Choi DH (2020) Federated reinforcement learning for energy management of multiple smart homes with distributed energy resources. IEEE Trans Industr Inf 18(1):488–497

Lee S, Ryu K (2022) Development of the architecture and reconfiguration methods for the smart, self-reconfigurable manufacturing system. Appl Sci 12(10):5172

Leng J, Liu Q, Ye S et al (2020) Digital twin-driven rapid reconfiguration of the automated manufacturing system via an open architecture model. Robot Comput-Integr Manuf 63:101895

Leng J, Wang D, Shen W et al (2021) Digital twins-based smart manufacturing system design in Industry 4.0: a review. J Manuf Syst 60:119–137

Li Y (2017) Deep reinforcement learning: an overview. arXiv:1701.07274

Liu C, Whitzer M, Yim M (2019) A distributed reconfiguration planning algorithm for modular robots. IEEE Robot Autom Lett 4(4):4231–4238

Liu R, Xie X (2024) Improve the industrial digital transformation through Industrial Internet platforms. Front Eng Manag 1–8

Ma Y, Du G, Jiao RJ (2020) Optimal crowdsourcing contracting for reconfigurable process planning in open manufacturing: a bilevel coordinated optimization approach. Int J Prod Econ 228:107884

Morgan J, Halton M, Qiao Y et al (2021) Industry 4.0 smart reconfigurable manufacturing machines. J Manuf Syst 59:481–506

Qin Z, Lu Y (2021) Self-organizing manufacturing network: a paradigm towards smart manufacturing in mass personalization. J Manuf Syst 60:35–47

Sutton RS, Barto AG (2018) Reinforcement learning: an introduction. MIT Press

Tang H, Li D, Wan J et al (2019) A reconfigurable method for intelligent manufacturing based on industrial cloud and edge intelligence. IEEE Internet Things J 7(5):4248–4259

Tao F, Zhang H, Liu A et al (2018) Digital twin in industry: state-of-the-art. IEEE Trans Industr Inf 15(4):2405–2415

Touzout FA, Benyoucef L (2019) Multi-objective sustainable process plan generation in a reconfigurable manufacturing environment: exact and adapted evolutionary approaches. Int J Prod Res 57(8):2531–2547

Wan J, Tang S, Li D et al (2018) Reconfigurable smart factory for drug packing in healthcare Industry 4.0. IEEE Trans Ind Inform 15(1):507–516

Wang J, Jiang C, Zhang K et al (2019) Distributed Q-learning aided heterogeneous network association for energy-efficient IIoT. IEEE Trans Industr Inf 16(4):2756–2764

Watkins CJCH, Dayan P (1992) Q-learning. Mach Learn 8:279–292

Yang S, Xu Z (2022) Intelligent scheduling and reconfiguration via deep reinforcement learning in smart manufacturing. Int J Prod Res 60(16):4936–4953

Yang J, Son YH, Lee D, Noh SD (2022) Digital twin-based integrated assessment of flexible and reconfigurable automotive part production lines. Machines 10:75
Yelles-Chaouche AR, Gurevsky E, Brahimi N et al (2021) Reconfigurable manufacturing systems from an optimisation perspective: a focused review of literature. Int J Prod Res 59(21):6400–6418
Yin YH, Xie JY, Da Xu L et al (2012) Imaginal thinking-based human-machine design methodology for the configuration of reconfigurable machine tools. IEEE Trans Industr Inf 8(3):659–668
Zhang C, Xu W, Liu J et al (2019) A reconfigurable modeling approach for digital twin-based manufacturing system. Procedia CIRP 83:118–125
Zhang C, Xu W, Liu J et al (2021) Digital twin-enabled reconfigurable modeling for smart manufacturing systems. Int J Comput Integr Manuf 34(7–8):709–733
Zheng P, Sang Z, Zhong RY et al (2018) Smart manufacturing systems for Industry 4.0: Conceptual framework, scenarios, and future perspectives. Front Mech Eng 13(2):137–150
Zhou X, Xu X, Liang W et al (2021) Intelligent small object detection for digital twin in smart manufacturing with industrial cyber-physical systems. IEEE Trans Industr Inf 18(2):1377–1386

Enhancing Resilience in Reconfigurable Manufacturing Systems Through Integrated Production and Maintenance Planning

Xingyu Li and Aydin Nassehi

Abstract Reconfigurable manufacturing systems (RMS) are designed to enhance adaptability in response to individualized customer demands, offering a promising pathway toward mass personalization. However, like traditional manufacturing systems, RMSs are susceptible to machine or module failures, which can hinder specific functionalities across interconnected machines and disrupt production flow. With the advancement of smart manufacturing technologies—such as embedded sensing and the Industrial Internet of Things—it is increasingly feasible to monitor machine conditions and implement preventive maintenance strategies to mitigate the risk of unexpected downtime. However, miscoordination between maintenance and reconfiguration activities can undermine system responsiveness and compromise its reconfigurability. In this context, striking an optimal balance between system adaptability and healthiness in delivering production demands remains a critical yet underexplored challenge. This chapter proposes a degradation-aware RMS decision-making model to optimally determine and adjust operational actions in real-time considering demand fulfillment, maintenance cost, and system health.

1 Introduction

This chapter introduces a novel approach for incorporating module degradation into the configuration optimization and reconfiguration decision-making of reconfigurable manufacturing systems (RMSs), which is adapted from the previously published work (Li et al. 2019). As customer demand becomes increasingly volatile and individualized, manufacturing systems face growing pressure to enhance flexibility while minimizing operational costs (Prasad and Jayswal 2019; Li et al. 2025).

X. Li (✉)
School of Engineering Technology, Purdue University, West Lafayette, IN, USA
e-mail: li4558@purdue.edu

A. Nassehi
School of Electrical, Electronic, and Mechanical Engineering, University of Bristol, Bristol, UK

S. Huang et al. (eds.), *Design and Operation of Smart Reconfigurable Manufacturing Systems in Industry 4.0/5.0*, Springer Series in Advanced Manufacturing,
https://doi.org/10.1007/978-3-032-00284-6_5

RMSs integrate the advantages of flexible manufacturing systems and dedicated manufacturing systems, offering both high throughput and adaptability to meet dynamic production requirements (Koren et al. 1999). These systems are composed of modular machines, where each machine consists of discrete modules that define its functionality and capacity. By reconfiguring the arrangement or substituting specific modules, RMSs can swiftly adjust to shifting production requirements.

Effective operation of RMSs depends on two fundamental decision-making components: identifying optimal subsequent configurations (Koren and Shpitalni 2010) and determining the appropriate timing and conditions for triggering system reconfigurations (Huang et al. 2018; Koren et al. 2018). Traditionally, the configuration design in RMS assumes that all modules are in a healthy operational state; module degradation is typically overlooked. Each module within an RMS may exhibit varying operational conditions—such as functional failure or compromised performance. A module failure can compromise the functionality of the associated machine and, in turn, degrade the overall system performance compared to its intended configuration. Therefore, the next-configuration design should not only aim for immediate performance optimization but must also strategically incorporate the current and projected health of individual modules to ensure sustained operational capability despite potential degradation.

Maintenance decisions introduce complex trade-offs for reconfiguration timing. While maintenance actions can enhance machine health and extend the reliability of future configurations, they also result in temporary machine unavailability, reducing the system's operational capacity in the short term. This dual impact creates a critical challenge: how to strategically determine the timing for system reconfiguration in light of changing demands and evolving module conditions (Ye et al. 2025). Addressing this challenge requires a coordinated decision-making framework that balances short-term functionality with long-term system adaptability and resilience (Ye et al. 2025; Gu et al. 2015).

Deep learning-based prognostic methods hold significant promise for interpreting real-time signals to assess module conditions and project near-future performance (Li et al. 2024). Recent studies have emphasized the importance of incorporating real-time condition data from system components (Vogl and Donmez 2015; Li et al. 2022a, b) to predict future system states (Li et al. 2018; Iijima and Takata 2016). Advances in multivariate analysis techniques (Matyas et al. 2017), cloud-based prognostics (Gao et al. 2015), and data-driven methods—especially those leveraging artificial intelligence (AI)—have shown strong potential in enabling accurate and timely maintenance decision-making (Cho et al. 2018). Such awareness of the manufacturing system is critical to proactively schedule maintenance according to the projected demands. However, there is still a lack of a decision-making framework that integrates all the factors together to make holistic decisions that balance the system's healthiness and responsiveness to the demands.

The work presented in this chapter aims to bridge the existing gap by proposing an integrated production and maintenance planning framework specifically designed for RMS. We present a decision-making framework that explicitly incorporates module degradation into the reconfiguration decision-making process, enhancing the system's robustness and resilience against potential module failures. Deep learning models are leveraged to enable intelligent, real-time assessment of evolving module conditions. A central challenge addressed in this work is the need to balance two competing actions: (1) reconfiguration actions that prioritize production efficiency and demand fulfillment, and (2) maintenance actions that seek to extend the operational life of machine modules. In the sections that follow, we introduce an integrated framework that incorporates module degradation into RMS operational planning. This is complemented by a case study inspired by real-world industrial settings, demonstrating the applicability and effectiveness of the proposed approach. Finally, the results are discussed, followed by concluding remarks and potential directions for future research.

2 Model Description

We consider a multi-stage RMS to satisfy the stochastically arriving demands while preserving system health. The inherent structural flexibility of RMSs enables dynamic, demand-responsive capacity provisioning through machine reconfiguration. We assume this reconfiguration action can contribute to accelerated module degradation due to increased usage, and also the reconfiguration action will lead to inherent system downtime. In the proposed model, module failure probabilities are modeled as functions of cumulative usage, while maintenance interventions are incorporated as restorative actions that mitigate degradation by reducing accumulated module usage, thereby prolonging module lifespans. Furthermore, module failure events—particularly when operation persists under low-health conditions—are modeled to impose prolonged recovery times via a repair action.

The integrated production and maintenance planning framework and the associated information flow are illustrated in Fig. 1. At each time step t, the system first assesses whether a reconfiguration—entailing a **re-mining of feasible configurations (Step A)**—is necessary, based on current demand realizations and the operational status of machines. If reconfiguration is not required, a set of **assembly, disassembly, relocation, and maintenance (ADLM) actions (Step B)** is scheduled to enable predictive maintenance or to repair malfunctioning modules. Conversely, if reconfiguration is warranted, the process begins with the **Generation of Configuration Candidates (Step C)**. Each candidate is subjected to a two-stage evaluation procedure:

- **Reconfiguration Planning (Step D)**: This step estimates the optimal operational condition achievable by each candidate, accounting for the current constraints on working time.
- **Capacity Distribution Estimation (Step E)**: This step projects the short-term probabilistic distribution of the system's capacity under each candidate configuration.

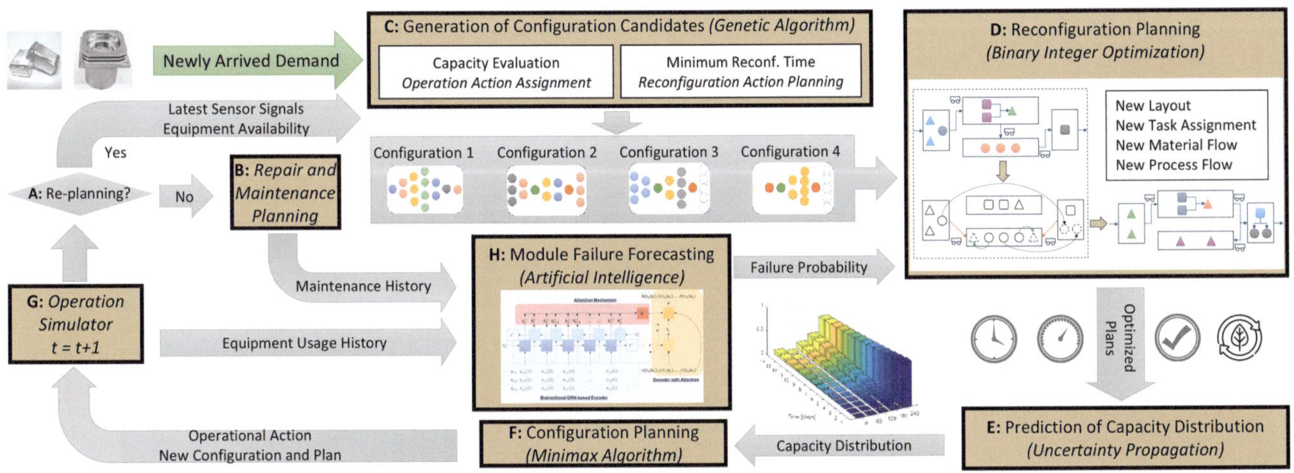

Fig. 1 The integrated production and maintenance planning framework for optimizing the robustness of RMS reconfiguration, adapted and extended from the previously published work (Li et al. 2019)

Based on the resulting capacity distributions, **Configuration Planning (Step F)** selects the candidate with the highest likelihood of satisfying demand as the system's new configuration. The selected operational actions are then passed to an **Operation Simulator (Step G)**, which updates both the historical action log and module usage history. These updated histories serve as input to a deep neural network model (Step H), which is continuously trained to update the estimated failure probabilities of each module based on cumulative usage and degradation patterns.

2.1 Reconfiguration, Repair and Maintenance Planning

The state of module i is represented by a binary vector $\boldsymbol{s}^i = [s_0^i, s_1^i, s_2^i, s_3^i, \ldots, s_{N_s}^i]$, where N_s is the number of machines. $s_k^i = 1$ indicates that module i is in machine k. Modules that are not in machines are in the inventory, denoted by machine $k = 0$. Module i can be removed from machine k and assembled in machine $k\prime$, , (operation action $o_{kk\prime}^i$). Machine-wise actions are also created to facilitate operations. Machine k, it can be relocated to the location of machine $k\prime$ (operation action $o_{k\prime}^k$) or it can be disassembled to inventory (operation action o_0^k). To increase the fidelity of the model, regular and quick actions (denoted by $\boldsymbol{q}$) are considered. Quick actions require less time to accomplish but degrade the module more. Module failure probability can be reduced by p_d^i after disassembly or by p_a^i by maintenance before assembly. Denote a new configuration by $\boldsymbol{S} = \left[\boldsymbol{s}^1, \boldsymbol{s}^2, \boldsymbol{s}^3, \ldots, \boldsymbol{s}^{N_m}\right]$, and aggregate all operation actions in vector $\boldsymbol{O}$. Configuration $\boldsymbol{S}$ can be calculated as $\boldsymbol{S} = \boldsymbol{BO}$, where matrix $\boldsymbol{B}$ records changes to the state of the configuration by operational actions. Operational actions cannot overdraw modules from the current configuration, $\boldsymbol{S}_0 - \boldsymbol{B}_0\boldsymbol{O} \geq 0$ (subscript 0 indicates the current configuration).

The degradation of machine k is characterized by the failure probability of machine k, which is estimated as the maximum module failure probability over all modules forming the machine. The failure probability of machine k is expressed as

$$u^k = \max_i \left(P_{di} s_k^i - P_{pi} p_a^i\right), \tag{1}$$

where P_{di} are module failure probabilities, and P_{pi} are reductions in module failure probabilities by maintenance. The optimization model is formulated:

$$\begin{aligned} \min_{\boldsymbol{o},\, \boldsymbol{p}, \boldsymbol{q}} \quad & \sum_{k=1}^{k=N_s} w_{dk} u^k + \sum_{i=1}^{i=N_m} \boldsymbol{c}_{oi}^T \boldsymbol{o}^i + \sum_{i=1}^{i=N_m} \boldsymbol{c}_{qi}^T \boldsymbol{q}^i + \sum_{i=1}^{i=N_m} \boldsymbol{c}_{pi}^T \boldsymbol{p}^i \\ \text{s.t. } (a) \quad & \sum_{i=1}^{i=N_s} \boldsymbol{t}_{oi}^T \boldsymbol{o}^i + \sum_{i=1}^{i=N_m} \boldsymbol{t}_{qi}^T \boldsymbol{q}^i + \sum_{i=1}^{i=N_m} \boldsymbol{t}_{pi}^T \boldsymbol{p}^i \leq \bar{t} \\ (b) \quad & \boldsymbol{S}_0 - \boldsymbol{B}_0 \boldsymbol{O} \geq 0, \boldsymbol{M}_{\boldsymbol{v}} \boldsymbol{S} = f_{mr}(\boldsymbol{v}, \boldsymbol{n}) \end{aligned} \tag{2}$$

$$(c)\ \boldsymbol{o}, \boldsymbol{q}, \boldsymbol{p} \text{are binary variables,}$$

where $\bar{t}$ is the available time, w_{dk} is the importance of machine k (which is related to the number of machines at the same stage as machine k), $\boldsymbol{c}_x$ denotes the increase in module failure probability due to action x, and $\boldsymbol{t}_x$ denotes the time for action x. $\boldsymbol{M}_v$ is a binary matrix that defines the modules in each machine. f_{mr} calculates the modules required for all machines in a configuration given by machine type $\boldsymbol{v}$ and number $\boldsymbol{n}$ of machines at each stage. Constraint (a) ensures that the ADLM actions are done in the available time; (b) ensures that the scheduled actions can be done with the available resources. The model represents the three main operations of interest in the reconfigurable manufacturing system: maintenance ($\boldsymbol{v} = \boldsymbol{v}_0, \boldsymbol{n} = \boldsymbol{n}_0$), repair ($\boldsymbol{v} = \boldsymbol{v}_0, \boldsymbol{n} > \boldsymbol{n}_0$) and re-configuration ($\boldsymbol{v} \neq \boldsymbol{v}_0, \boldsymbol{n} \neq \boldsymbol{n}_0$) where $\boldsymbol{v}, \boldsymbol{n}$ are machine type and number at all stages, and subscript 0 indicates current time. Maintenance is triggered once spare time is available. Reconfiguration and repair decisions are triggered by events such as unfulfilled demands and module failure. Reconfigurations require the new configuration to be determined before planning for operational actions.

2.2 Generation of Candidate Configurations

To generate candidate configurations, the approach of the authors in Ye et al. (2025) is modified to generate configurations $(\boldsymbol{v}, \boldsymbol{n})$ with large capacity that have also low module failure probability, i.e.,

$$\max_{\boldsymbol{v},\boldsymbol{n}} \sum_{k=1}^{k=N_p} w_{pk} f_{pr}(k, \boldsymbol{v}, \boldsymbol{n}) - w_t f_{rt}(\boldsymbol{v}, \boldsymbol{n}, \boldsymbol{v}_0, \boldsymbol{n}_0) - w_p f_p(\boldsymbol{v}, \boldsymbol{n}, \boldsymbol{P}_d)$$

$$\text{s.t. } (a)\ f_{m_p}(\boldsymbol{v}, \boldsymbol{n}) \leq s_{m_p}(t), \forall p \tag{3}$$

$$(b)\ 1 \leq v_{t_s} \leq N_{sv}, 0 \leq n_{t_s} \leq N_{sn} \text{and integer}, \forall t_s,$$

where f_{pr} represents an optimization model that calculates the maximum capacity for part k, f_{rt} is an optimization model that evaluates the minimum reconfiguration time ignoring module degradation, f_p is a function that estimates the overall failure probability of a configuration. Weight factors w_x are used to add preferences to the configuration candidates, i.e., higher capacity for some of the parts, quicker reconfiguration or healthier configuration. The approach generates a population of candidate configurations $\boldsymbol{C} = \left[\boldsymbol{c}^1, \boldsymbol{c}^2, \ldots, \boldsymbol{c}^{N_{cd}}\right] = \left[\left(\boldsymbol{v}^1, \boldsymbol{n}^1\right), \ldots, \left(\boldsymbol{v}^{N_{cd}}, \boldsymbol{n}^{N_{cd}}\right)\right]$, which are used in making reconfiguration decisions.

2.3 Evolution of Capacity Distribution: Uncertainty Propagation

We consider a configuration c_0 with initial capacity pr_0. When a module of a machine fails, that machine fails, and configuration c_0 is damaged. All the remaining undamaged machines in configuration c_0 can be thought of as another configuration c_i. Without repair or reconfiguration, configuration c_i continues production until one of these modules fails. This process is repeated until the system completely loses its capacity. A robust configuration is one where the probability of satisfying the demand is high even when some modules fail. Robustness is critical once unexpected damage occurs, and demand is urgent.

The number N_d of possible damaged configurations that can result from one single healthy configuration is $2^{N_s} - 1$ which makes exhaustive analysis of all damaged configurations intractable. Thus, the evolution of capacity is used as an alternative approach. Considering that damage can only lower the capacity of a configuration, the capacity evolution over time is

$$\begin{bmatrix} \mathrm{P}(pr_0(t+1)) \\ \mathrm{P}(pr_1(t+1)) \\ \mathrm{P}(pr_2(t+1)) \\ \vdots \\ \mathrm{P}(pr_N(t+1)) \end{bmatrix} = \begin{bmatrix} A_{00} & 0 & 0 & & 0 \\ A_{01} & A_{11} & 0 & \cdots & 0 \\ A_{02} & A_{12} & A_{22} & & 0 \\ & \vdots & & \ddots & \vdots \\ A_{0N} & A_{1N} & A_{2N} & \cdots & 1 \end{bmatrix} \begin{bmatrix} \mathrm{P}(pr_0(t)) \\ \mathrm{P}(pr_1(t)) \\ \mathrm{P}(pr_2(t)) \\ \vdots \\ \mathrm{P}(pr_N(t)) \end{bmatrix}, \tag{4}$$

where $P(x)$ represents the probability that the RMS capacity is x, $pr_i(t)$ decreases as i increases, N represents the number of different capacities of all configurations, and $pr_N(t)$=0. It is assumed that module failures are independent stochastic variables. Coefficients A_{ij} are obtained as follows:

1: Compute original and damaged configuration capacities by f_{pr}.
2: Cluster all configurations by capacity.
3: For each configuration c_i (with n_{t_s} machines at stage t_s and non-zero capacity), compute the probability $P_{c_i c_j}$ that one of its modules will fail and become configuration c_j (with n_{m,t_s} machines at stage t_s and non-zeros capacity) by using

$$P_{c_i c_j} = \prod_{t_s=1}^{t_s=N_s} \sum_{k}^{n_{t_s}} (1 - p_{dk})^{(1-I_{dk})} (p_{dk})^{(I_{dk})}, \tag{5}$$

where I_{dk} is a binary value that records if machine k is damaged (1) or not (0), and hence $\sum_{k=1}^{k=n_{t_s}} I_{dk} = n_{m,t_s}$, and p_{dk} is the failure probability for machine k, calculated as

$$p_{dk} = 1 - \prod_{i=1}^{i=m_{v_k}} \left(1 - p_{m_i}\right), \tag{6}$$

where m_{v_k} is the number of modules in machine k, and p_{m_i} is the failure probability of module i

4: Compute coefficients A_{ij} as

$$A_{ij} = p_{pr_i pr_j} = \sum_{i}^{N_c} \sum_{j \neq i}^{N_c} I_{c_i c_j} p_{c_i c_j}, \forall i \in [1, N], \forall j, \tag{7}$$

$$A_{0j} = p_{pr_0 pr_j} = 1 - \sum_{j=1}^{j=N-1} p_{pr_0 pr_j}, \forall j, \tag{8}$$

where $I_{c_i c_j} = 1$ if $pr_i = f_{pr}(\boldsymbol{c_i})$ & $pr_j = f_{pr}(\boldsymbol{c_j})$, and 0 otherwise.

Capacities pr_i can be grouped in a vector $\boldsymbol{pr}$. Starting from a configuration with capacity pr_0, the probabilities of available capacity to be the entries in $\boldsymbol{pr}$ over a time horizon $t = 0, \ldots, h$ are

$$\mathrm{P}(\boldsymbol{pr}(t+h)) = \boldsymbol{A}^h \mathrm{P}(\boldsymbol{pr}(t)) = \boldsymbol{A}^h [100 \ldots 0]^T. \tag{9}$$

2.4 Configuration Planning

Given the evolution of capacity probability over time, the performance of a candidate configuration is measured by the probability of satisfying the demand received in the planning horizon. First, the processing time $pt_k(t)$ for each part k is calculated. The number of parts k processed between time t and $t+1$ is $\phi_k(t) = pr_k(t) pt_k(t)$. The variance of $\phi_k(t)$ is given by

$$\mathrm{Var}[\phi_k(t)] = \mathrm{Var}\left[pr_k(t) pt_k(t)\right] = pt_k^2(t) \mathrm{Var}\left[pr_k(t)\right], \tag{10}$$

$$\mathrm{Var}\left[\sum_{\tau=t+1}^{\tau=t+t_p} pr_k(\tau) pt_k(\tau)\right] = \boldsymbol{pt}_k^T \boldsymbol{\Sigma^k} \boldsymbol{pt}_k \tag{11}$$

where $\boldsymbol{\Sigma^k}$ is the capacity covariance matrix with entries given by $\Sigma_{ij}^k = cov\left(pr_k(\tau),\ pr_k\left(\tau'\right)\right) = \mathrm{E}\left(pr_k(\tau)\ pr_k\left(\tau'\right)\right) - \overline{pr_k}(\tau)\ \overline{pr_k}\left(\tau'\right)$.

The processing time can be computed by minimizing $\mathrm{Var}[\phi_k(t)]$:

$$\min_{\boldsymbol{pt}} \quad \sum_{k} \boldsymbol{pt}^{T} \boldsymbol{\Sigma}^{k} \boldsymbol{pt} \tag{12}$$

$$\text{s.t. } (a) \sum_{\tau=t+1}^{t_{n_d}} \overline{pr_k}(\tau) pt_k(\tau) \geq \sum_{\tau=t+1}^{t_{n_d}} d_k(\tau), \forall n_d, \forall k$$

$$(b) \left[\sum_{k} pt_k(\tau) + rt(\tau) \right] \leq \bar{t} - rt(\tau), \quad pt_k(\tau) \geq 0 \ \forall \tau,$$

where constraint (a) ensures that all demands received can be satisfied by the average capacity, and constraint (b) ensures that the total processing is done in the available time. As uncertainty increases with time, the model prefers to process the parts early in the planning horizon. The vector of processing times is updated at each time step according to the new demand received and according to the current system status.

The configuration with the lowest failure probability is selected, by solving the following binary integer-programming problem

$$\min_{z} \quad \sum_{k} \sum_{n_d} \alpha_{k,n_d} \left(\boldsymbol{c}^{z} \right), \tag{13}$$

where $\boldsymbol{c}^{z} \in \boldsymbol{C}$, and α_{k,n_d} describes the probability of marginal failure for satisfying demand n_d for part k, which follows conditions

$$\mathrm{P}\left(\sum_{\tau=t+1}^{t_{n_d}} pr_k(\tau) pt_k(\tau) \geq \sum_{\tau=t+1}^{t_{n_d}} d_k(\tau) \right) \geq 1 - \alpha_{k,n_d} \tag{14}$$

The total demand before the due time is $\lambda_{k,n_d} = \sum_{\tau=t+1}^{t_{n_d}} d_k(\tau)$, and the total production before the due time is $X_{k,n_d} = \sum_{\tau=t+1}^{t_{n_d}} pr_k(\tau) pt_k(\tau)$. Because X_{k,n_d} follows a non-Gaussian distribution, Cantelli's inequality and Eq. (14) are used to estimate α_{k,n_d}, which is a metric for configuration performance:

$$\mathrm{P}(X - \mathrm{E}[X] \geq \lambda) \geq \frac{\sigma^2}{\sigma^2 + \lambda^2}, \text{ and thus } \alpha_{k,n_d} = \frac{{\lambda_{k,n_d}}^2}{\sigma_X^2 + {\lambda_{k,n_d}}^2}. \tag{15}$$

By using this model, the selected configuration will not only satisfy the demand but will also very likely satisfy the demand even when some of its modules fail. This robustness is important for the RMS operation, as the occurrence of damage is usually unpredictable and performing reconfigurations may be ineffective due to time pressure. Following this model, reconfiguration decisions are triggered once one of the following conditions holds: (1) the current configuration cannot satisfy the new demand; (2) the module stocks are changed.

2.5 Module Failure Forecasting

An accurate estimation of failure probability is essential for describing the ability of the module to function under given conditions for a specified period of time. The diversity in reconfiguration actions raises the difficulty of identifying the causes of module failures. The sequence of operational actions also affects the chances of failure, i.e., accumulation of failure probability depends on the sequence of usage the module has experienced. To address this, a data-driven approach is used to estimate the module failure probability by using a deep-learning model with long short-term memory (LSTM). LSTM is superior to other models due to its strength of capturing the dynamic behavior for a time sequence. The model created in this study contains a sequential input layer, an LSTM layer, a hidden layer (with 25 nodes), and a regression layer. For training, all operational actions exerted are recorded as a 'resumé' of the module, including the action type and time. Actions are discretized and recorded as a matrix of binary vectors. The output of the training data is the estimated usage of each module, which is assumed to be measurable. With continuous retraining by the data from daily operation, the model can be calibrated and used for forecasting the damage probability given the resumé of each module.

3 Prototype Implementation

The example from previous work of the authors (Iijima and Takata 2016) is used to test a prototype implementation of the model. A company invests in a single processing line with a limited number of modules. Two types of parts with multiple features are to be processed. 5 types of machines with multiple variants are available for the processing service. The modules stochastically get damaged based on their usage history. Once damaged, module refills are ordered immediately and arrive in 14 days. The assembly time is 0.2 h for machine tools and 1 h for bases or arms. Disassembly times are half of assembly times. The time durations of individual actions are as follows: relocation among stages requires 0.2 h; relocation between a stage and storage takes 0.5 h; and maintenance activities require 4 h. The total time for ADLM operations is computed as the sum of the durations of the respective actions.

4 Results

A case study is performed to illustrate the interconnections between the reconfiguration and maintenance actions that will impact the responsiveness to the market demands and the evolving module conditions. Specifically, we vary the total available operational time $\bar{t}$ and analyze its impact on the allocation of time across different action categories, including regular versus expedited reconfiguration and various maintenance-related interventions. The results of this analysis are presented in Fig. 2.

A minimum of six hours of available time is required to complete a system reconfiguration. As the available time increases, the integrated production and maintenance planning framework shifts from favoring quick actions to regular reconfiguration processes, thereby mitigating excessive module degradation associated with rapid reconfiguration. When the available time exceeds 33 h, the model begins to incorporate maintenance actions specifically targeting the health improvement of modules held in storage. A comparative analysis of module degradation across system stages reveals that, although a new configuration can technically be implemented within a six-hour threshold, its capacity to support sustained, high-performance operation is significantly limited due to compromised module conditions resulting from the absence of maintenance.

Figure 3a illustrates the effect of available working time on the average failure probability across the entire manufacturing system. It is evident that as working time increases, the average failure probability decreases, primarily due to the incorporation of maintenance activities. In Fig. 3b, the scenario with a relatively long working time (36 h) and a scheduled maintenance plan is presented. This figure demonstrates the relationship between the number of machines at each stage and the corresponding average module failure probability. Notably, the proposed framework strategically assigns higher-risk machines to stages with a larger number of machines. This approach ensures that, even in the event of module failure at these stages, alternative machines remain available, thereby reducing the risk of total system capability loss and enhancing overall system resilience.

Figure 4 illustrates the significant impact of system reconfiguration on mitigating degradation in module condition under varying demand pressures. Prior to each reconfiguration event, the system experiences a marked rise in averaged failure probability, closely aligned with surges in unfulfilled demands, indicating escalating production stress. The first system reconfiguration, implemented around time

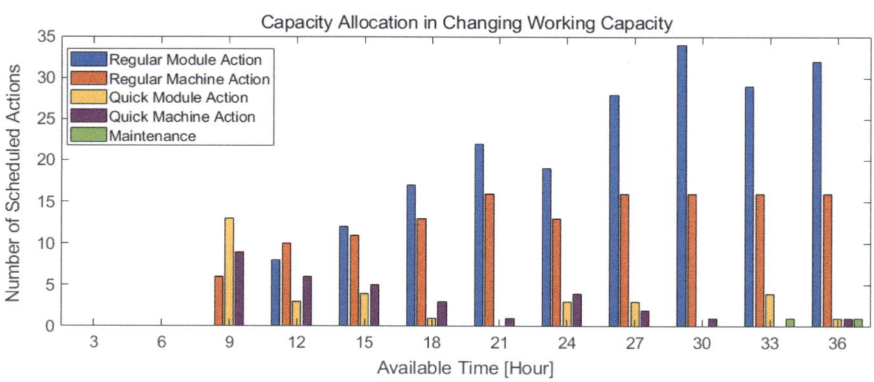

Fig. 2 Variation in action allocation with increasing available working time

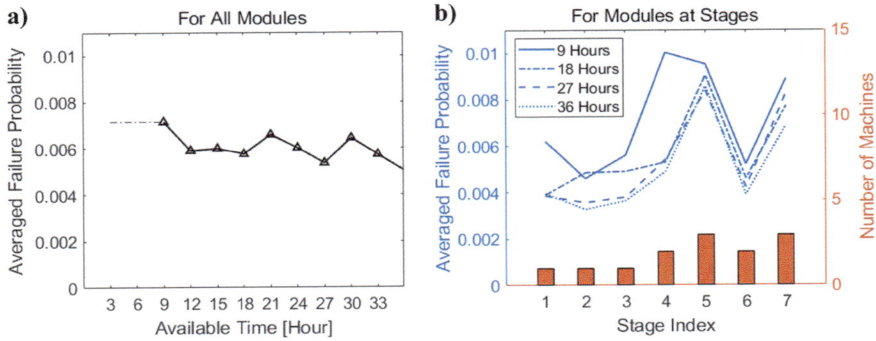

Fig. 3 Impact of available working time on module failure probabilities

Fig. 4 Impact of system reconfiguration on module reliability and demand fulfillment over time

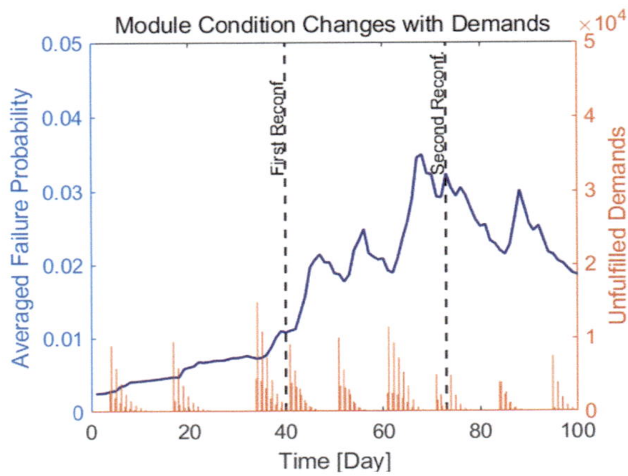

step 40, leads to an immediate reduction in failure probability, demonstrating its effectiveness in restoring partial system stability. However, as demands persist, the failure probability rises again, necessitating a second reconfiguration around time step 70. This second intervention once more curtails the upward trend in failure probability, suggesting that system reconfiguration plays a crucial role in interrupting deterioration cycles and enhancing system resilience under dynamic loading conditions.

According to Fig. 5, the two marked reconfiguration events have a pronounced effect on the system's resource availability, reflecting an enhancement in the overall RMS performance. The first reconfiguration, occurring around time step 40, leads to a noticeable increase in the availability of Type 1 and Type 4 modules, suggesting that the system underwent a structural simplification that reduced the dependency on these modules. Similarly, the second reconfiguration, near time step 70, results in a further rise in available modules, particularly for Types 1, 3, and 4, reinforcing the inference that each reconfiguration contributes to a more efficient allocation of

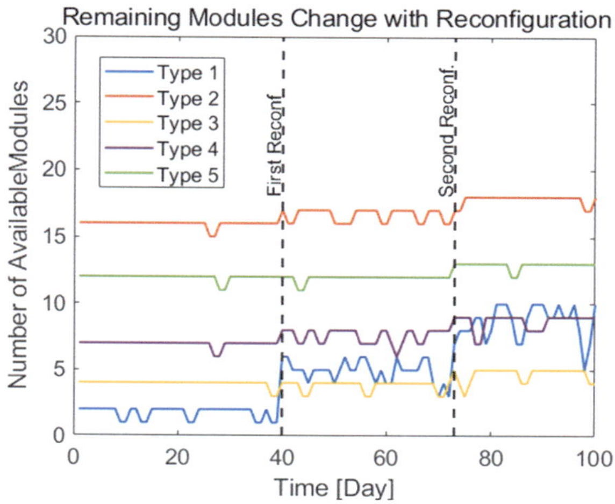

Fig. 5 Impact of system reconfiguration on module availability over time

module resources. These increases in module availability, without corresponding increases in demand, imply that the reconfigurations likely minimized redundancy and improved module utilization. Those additional modules thus act as spare capacity, enhancing the redundancy and resilience of the overall reconfigurable manufacturing system.

Figure 6 illustrates the time evolution of averaged failure probabilities for modules with and without system reconfiguration. In the scenario without reconfiguration (red lines), particularly for the most degraded 10% of modules (dashed red line), the failure probability rises sharply, peaking around Day 62—corresponding to the point where the configuration exhausts its operational useful life. This is marked by a red star, signaling module failure due to cumulative degradation and triggering a prolonged recovery period. In contrast, the system with reconfiguration (blue lines) maintains significantly lower failure probabilities throughout the time horizon. The top 10% degraded modules in this scenario (dashed blue line) experience a temporary increase in failure risk, but this is mitigated by timely reconfiguration events that redistribute the workload and reduce module stress. Consequently, the reconfigurable system avoids catastrophic failure, ensuring continuous operation and minimizing unmet demand—highlighting the value of dynamic reconfiguration in preserving RMS performance and enhancing system resilience.

Figure 7 illustrates the evolution of system capacity over time for two parts under scenarios with and without reconfiguration. The red curves (without reconfiguration) show abrupt and severe drops in capacity, with both Part 1 and Part 2 eventually dropping to zero capacity around Day 62. This failure aligns with previously observed peaks in failure probability and indicates total system failure due to module degradation and insufficient maintenance adaptability. In contrast, the blue curves (with reconfiguration) demonstrate a more stable and controlled capacity trajectory, albeit at a modestly reduced level. These reductions reflect the shift to simplified configurations that utilize fewer modules, as highlighted in Fig. 5. Despite operating with

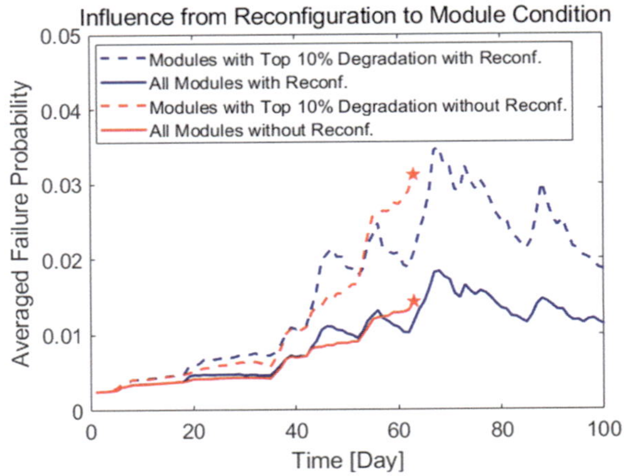

Fig. 6 Comparison of the distribution of module failure probability with and without reconfigurations

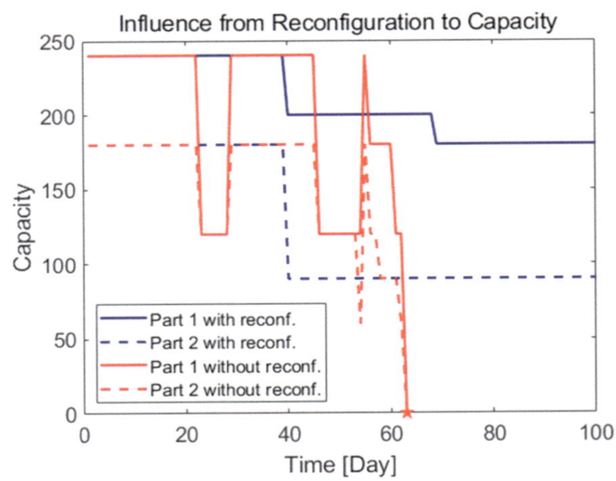

Fig. 7 Comparison of the system capacity with and without reconfigurations

lower capacity, the RMS sustains demand fulfillment while also benefiting from decreased module usage. This leads to less frequent and more manageable maintenance, which, over time, contributes to a decline in failure probability. Thus, the reconfigured system trades off a small loss in throughput for improved maintainability, higher module availability, and enhanced operational resilience—ultimately achieving more robust long-term performance compared to the non-reconfigurable scenario.

5 Conclusion

This chapter presented a novel degradation-aware reconfigurable manufacturing system. Our results demonstrate that incorporating the degradation of machine modules into the reconfiguration process enables the creation of resilient reconfigurable manufacturing systems capable of managing damaged modules while ensuring timely demand fulfillment and efficient responsiveness. This approach integrates an AI-driven predictive model of degradation with optimization strategies for both maintenance and reconfiguration to achieve these objectives. Currently, the model triggers reconfiguration reactively in response to shifts in demand or machine availability. Future research could focus on exploring proactive reconfiguration strategies and further investigating the deployment of this model in real-world production environments.

References

Cho S, May G, Tourkogiorgis I, Perez R, Lazaro O, Maza B, Kiritsis D (2018) A hybrid machine learning approach for predictive maintenance in smart factories. In: Conference on advances in production management systems, pp 311–317

Gao R, Wang L, Teti R, Dornfeld D, Kumara S, Mori M, Helu M (2015) Cloud-enabled prognosis for manufacturing. CIRP Ann 64(2):749–772

Gu X, Jin X, Ni J, Koren Y (2015) Manufacturing system design for resilience. Procedia CIRP 36:135–140

Huang S, Wang G, Shang X, Yan Y (2018) Reconfiguration point decision method based on dynamic complexity for reconfigurable manufacturing systems. J Intell Manuf 29(5):1031–1043

Iijima H, Takata S (2016) Condition based renewal and maintenance integrated planning. CIRP Ann 65(1):37–40

Koren Y, Shpitalni M (2010) Design of reconfigurable manufacturing systems. J Manuf Syst 29(4):130–141

Koren Y, Heisel U, Jovane F, Moriwaki T, Pritschow G, Ulsoy G, Van Brussel H (1999) Reconfigurable manufacturing systems. CIRP Ann 48(2):527–540

Koren Y, Gu X, Guo W (2018) Reconfigurable manufacturing systems. Front Mech Eng 13(2):121–136

Li X, Bayrak AE, Epureanu BI, Koren Y (2018) Real-time teaming of multiple reconfigurable manufacturing systems. CIRP Ann 67(1):437–440

Li X, Nassehi A, Epureanu BI (2019) Degradation-aware decision making in reconfigurable manufacturing systems. CIRP Ann 68(1):431–434

Li X, Koren Y, Epureanu BI (2022a) Complementary learning-team machines to enlighten and exploit human expertise. CIRP Ann 71(1):417–420

Li X, Krivtsov V, Arora K (2022b) Attention-based deep survival model for time series data. Reliab Eng Syst Saf 217:108033

Li X, Athinarayanan R, Wang B, Yuan W, Zhou Q, Jun M, Bravo J, Gao RX, Wang L, Koren Y (2024) Smart reconfigurable manufacturing: literature analysis. Procedia CIRP 121:43–48

Li X, Nassehi A, Hu SJ, Joung BG, Gao RX (2025) A large manufacturing decision model for human-centric decision-making. CIRP Ann

Matyas K, Nemeth T, Kovacs K, Glawar R (2017) A procedural approach for realizing prescriptive maintenance planning in manufacturing industries. CIRP Ann 66(1):461–464

Prasad D, Jayswal S (2019) A review on flexibility and reconfigurability in manufacturing systems, In: Innovations in materials science and engineering, pp 187–200. Springer, Singapore
Vogl GW, Donmez MA (2015) A defect-driven diagnostic method for machine tool spindles. CIRP Ann 64(1):377–380
Ye W, Yang S, Li X (2025) A bi-level supply chain resilience model using cloud manufacturing. J Manuf Syst 80:662–672

Autonomous Coordination of Multiple Reconfigurable Manufacturing Systems

Xingyu Li

Abstract Reconfigurable manufacturing has been designed to dynamically adjust the hardware, software, and logistics to rapidly match the system capability and functionality to the changing markets. The recent emergent mass personalization and mass individualization demands the future reconfigurable manufacturing systems (RMSs) with greater flexibility and adaptability, primarily due to the substantial increase in product variety and demand uncertainty. In this study, we investigate a manufacturing environment comprising multiple reconfigurable production lines that are designed to simultaneously produce a diverse product mix. These lines share modular components, enabling dynamic reconfiguration of each line in real time to adapt both capacity and functionality. Leveraging recent advances in smart manufacturing, we propose a real-time, autonomous coordination framework for RMSs. This framework enables autonomous production lines to coordinate collaboratively, enhancing the system's responsiveness and collective ability to meet fluctuating market requirements. Our approach encompasses detailed modeling to automate the operations of each line, including configuration design, capacity planning, reconfiguration action planning, processing time allocation, and configuration path optimization for each individual line. Furthermore, we incorporate real-time information sharing among the lines and introduce a mathematical framework to support the autonomous coordination of stochastic demand. This includes a bidding-based mechanism for distributing demand and exchanging machines or modules (i.e., groups of machines) among lines. The proposed method enables adaptive configuration of lines and machines in response to shared demand while operating under constraints related to the limited availability of configurable components.

X. Li (✉)
School of Engineering Technology, Purdue University, West Lafayette, IN, USA
e-mail: li4558@purdue.edu

S. Huang et al. (eds.), *Design and Operation of Smart Reconfigurable Manufacturing Systems in Industry 4.0/5.0*, Springer Series in Advanced Manufacturing,
https://doi.org/10.1007/978-3-032-00284-6_6

1 Introduction

This chapter introduces a novel approach for autonomous coordination of the manufacturing resources and demands among multiple reconfigurable manufacturing systems (RMSs), which is adapted from the previously published work (Li et al. 2018). Unpredictable and frequent market changes, including demand fluctuations and rapid introduction of new products, make the development of cost-effective and responsive manufacturing systems a necessity for companies to survive in the new competitive environment. RMSs, which combine high throughput from dedicated manufacturing lines and flexibility from flexible manufacturing systems, have been introduced to address this challenge (Koren et al. 1999). Flexibility in the RMS structure not only enhances the RMS responsiveness through upgradable capacity and modifiable functionality but also provides the opportunity to boost the utility of machines and the entire manufacturing system by using the exact functionality and capacity once needed (Koren et al. 2017). To dynamically match RMS configurations to demands, efficient and effective real-time management strategies are indispensable for RMS to guide system reconfiguration, not only for the single machine functionality adjustments but also for coordinating multiple resources, including both machines and modules, to be placed in the effective layout to dynamically streamline the process and material flows.

There is a rich literature on RMS planning, including process (Bensmaine et al. 2014; Azab and ElMaraghy 2007), capacity (Asl and Ulsoy 2003; Ceryan and Koren 2009; Epureanu et al. 2021), and configuration planning. RMS configuration planning requires a detailed consideration of operational actions. Wang and Koren (2012) investigated capacity planning by adjusting the RMS configuration to optimize the system throughput. Youssef and ElMaraghy (2007) proposed an optimal configuration selection strategy to customize RMS configurations to different demand scenarios by optimizing the similarity between the adjacent configurations to increase the system utilization.

Previous studies that consider the stochastic nature of market demands assume that the distribution of demand over time is deterministic and the reconfiguration occurs within a single RMS. However, emerging demands in personalized manufacturing (Gu and Koren 2022) and remanufacturing (Kim et al. 2024) increase the difficulty of demand forecasting and the diversity of products, which limits the application of existing RMS solutions. Moreover, the growing prevalence of mixed product portfolios—fueled by heightened consumer expectations for customization and accelerated product life cycles—has introduced greater variability and fragmentation into production environments. In light of these challenges, there is a pressing need for intelligent, adaptive solutions that go beyond the management of individual RMSs. These solutions must address the increasing demand for inter-system collaboration, enabling dynamic allocation of shared resources, synchronized reconfiguration strategies, and collective responsiveness to evolving market conditions (Li et al. 2024a, b).

The advent of autonomous systems in manufacturing has enabled automated guided vehicles (AGVs) and autonomous mobile robots to independently navigate and manage material handling tasks, thereby improving the efficiency of module allocation and part flow balancing among workstations. Equipped with advanced environmental perception capabilities, these systems offer the potential to significantly reduce reconfiguration time and effort. Inoue et al. introduced a novel approach to RMSs that incorporates autonomous mobile manipulators—not only for material transport and handling, but also for directly modifying manufacturing processes and system configurations, such as adding, removing, or replacing machines (Inoue et al. 2021). Terzić et al. further conceptualize decision-making functions as modular components that can be flexibly adapted to diverse resource configurations (Terzic et al. 2009). Li et al. developed the human-centric adaptive manufacturing that considers the autonomous resources collaboratively executing manufacturing operations, interacting intelligently with their environments, and coordinating with other agents and human operators to enable flexible and distributed system reconfiguration across multiple locations for both the purpose of the production and education (Li et al. 2023, 2025).

Most of the studies utilize the AGVs as mobile machines to automate the material flows between the reconfigurable machines or reconfigurable manufacturing workstations. However, there remains a significant gap in the literature regarding the use of AGVs to coordinate operation of multiple RMSs toward a shared goal. Specifically, given the limited manufacturing resources and emergent demands, how to dynamically allocate them to separate lines to maximize the demand fulfillment rate. This requires a clear understanding of the potential capability improvements achievable by each line with additional resources, as well as the associated costs and efforts required for reconfiguration.

In this study, we propose a multiple RMS coordination framework that synergistically integrates modularity and autonomy by conceptualizing manufacturing lines as dynamic groups of reconfigurable machines capable of sharing resources and executing self-directed reconfiguration. In addition to the fundamental features of traditional RMSs [2], the proposed system incorporates two key attributes: autonomy (in the line-level reconfiguration) and coordination, through autonomous coordination for demand and resource sharing at the factory level. To realize this vision, we introduce a real-time decentralized management architecture in which reconfigurable machines are organized into production lines that can flexibly reconfigure based on demands and available resources. An agent-based decision-making framework supports the coordination of lines, enabling the exchange of resources and demands across lines via a bidding algorithm.

2 Problem Statement

We consider a modern factory composed of multiple autonomous RMSs designed to produce a mix of part types $k \in K$. Each part contains diverse features and requires sequential operations. We assume that each line consists of a multi-stage RMS and the associated configurable components, i.e., modules. Each machine can be reconfigured by swapping its modules to adjust the machine's functionality and capacity. We refer to the machines in different configurations as variants. The system-level configuration is characterized by changes in the system layout, including the number and types of machine variants deployed at each stage, as well as the associated operations required to sequentially satisfy the process requirements of the parts. When considering multiple RMS lines, each addressing the demand of a specific part type, two key challenges arise: (1) as the number of possible RMS configurations grows exponentially with the complexity of the line (Koren and Shpitalni 2010), how can one efficiently determine the optimal configuration given limited resources to meet the required demand; and (2) how can resources and part demands be effectively allocated across lines to ensure a better match between line capabilities and assigned tasks, thereby improving overall demand fulfillment.

In this study, we address the challenge of dynamically reconfiguring multiple RMSs in response to changing demand. Due to the inherent complexity of the problem, centralized management strategies are often inefficient for identifying near-optimal resource and demand allocation plans across multiple lines. Furthermore, the number of relocation decisions increases rapidly with the growing number of production lines and resource types. When combined with the planning complexity inherent in RMSs, such challenges render existing optimization tools inadequate for practical implementation. To overcome this limitation, we propose a novel decentralized framework in which each production line is modeled as an intelligent agent capable of making autonomous reconfiguration decisions. These agents are also able to negotiate with one another to exchange modules and balance demand, thereby enabling more flexible, scalable, and adaptive system-wide coordination.

3 Model Description

In this section, we develop a two-stage decision-making framework that enables the dynamic and collaborative operation of multiple RMSs. The structure of the proposed system is illustrated in Fig. 1. The first stage introduces a bidding-based coordination algorithm that facilitates the reallocation of both production demands and manufacturing resources among RMS lines. Each line begins by analyzing its current set of resources and demands. Based on this internal assessment—which involves tasks such as demand potential assessment, operation-resource assignment, capacity planning, and processing path planning—the line determines its ability to fulfill assigned demands. The outcome of this planning yields a set of remaining

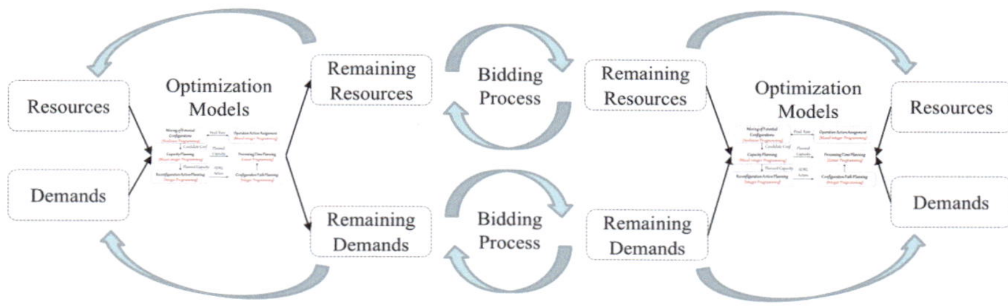

Fig. 1 Bidding and reconfiguration framework for multi-line RMS coordination

demands and remaining resources, which are then made available for sharing across lines.

In the second stage, the remaining resources and demands from one line are communicated and considered by adjacent lines in a peer-to-peer manner. Each RMS line then integrates the received resources and demands into its own planning framework. This repeated exchange and integration process continues until a stable allocation is achieved or a termination condition is met. During each iteration, autonomous decision-making within each line includes reconfiguration cost planning and configuration path recovery to determine the most efficient use of newly acquired resources or reassigned tasks.

3.1 Resource and Demand Reallocation Among Lines

Given limited module inventories and fluctuating demands, effective task and resource allocations are important to maintain a high processing efficiency and resource utility rate. To facilitate resource and demand sharing while keeping the operations and configuration of each line effective, we choose to model each individual RMS (i.e., each line) as an intelligent agent, which negotiates with others if the line cannot fulfill the received demand alone. Denote by f_u the unfulfilled orders (details in Sect. 3.7). We propose a structured multi-step bidding process, detailed in Algorithm 1.

Algorithm 1: Multi-step Bidding Process for Demand and Resource Reallocation.

1. **Initialize:** Distribute new received demands by heuristic rules, such as weighted/equal distribution.
2. **Tenderee Selection:** Each agent predicts its unfulfilled demands by using $\boldsymbol{f_u}$. The agent with the highest value is designated as the tenderee, and the other agents become bidders.
3. **Bidding Document:** The tenderee is responsible to provide a bidding document, outlining possible options to reduce unfulfilled demands, namely
   - A list of unachievable demands and currently available resources.
   - A request for additional modules to enhance its processing capacity.
4. **Bidding:** Each bidder evaluates the scenarios outlined in the bidding document and simulates reallocation outcomes. The scenario that results in the lowest overall unfulfilled demand for the system becomes the bidder's proposal to the tenderee.
5. **Evaluation:** The proposal resulting in the lowest total unfulfilled demand is selected. Corresponding demand and resources are reallocated between the tenderee and the winning bidder according to the selected scenario.
6. **Termination Check:** The iterative bidding process terminates if either of the following conditions is met:
   - The total unfulfilled demand across all agents remains unchanged from the previous iteration.
   - All demands are successfully fulfilled (i.e., total unfulfilled demand is zero).

Otherwise, the process returns to Step 2 for the next iteration.

The proposed algorithm is guaranteed to converge in a finite number of iterations, as the total volume of unfulfilled demand is finite and strictly decreases or remains unchanged in each iteration. Moreover, because the bidding evaluations are performed independently by each agent, parallel computation strategies can be effectively employed to enhance the scalability and efficiency of the algorithm. Following the completion of the bidding process and subsequent negotiation outcomes, each RMS performs real-time reconfiguration and task scheduling based on the newly assigned demands and resource allocations. The updated demands and resources are then computed accordingly to inform subsequent iterations or finalize the demand and resource allocations. Figure 2 shows an overview of the key decision-making steps involved in calculating the new updated remaining resources and demands, along with the associated mathematical programming models indicated in red within the brackets.

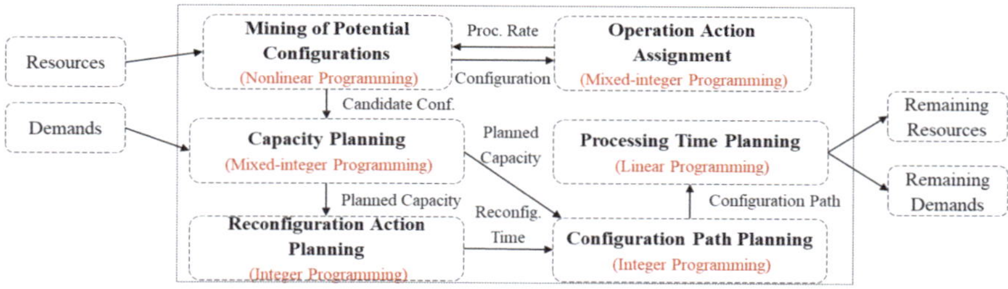

Fig. 2 A framework of mathematical models developed to plan the configuration of each individual RMS and compute the remaining demands and available resources

3.2 Operation Action Assignment

To customize an RMS so that it provides the exact capacity and functionality closely aligned with demand requirements (Koren et al. 1999), we find the marginal capacity of each potential configuration. At each stage t_s, we assume only one machine variant v_{t_s} is used. Given the processing operation op_{k,t_s} and the number of machines n_{t_s}, the processing rate of part k at stage t_s is $spr_{k,t_s} = \boldsymbol{f_c}(k, v_{t_s}, n_{t_s})\boldsymbol{op}_{k,t_s}$, where $\boldsymbol{f_c}$ is a vector function to calculate the stage capacity for all operations, $\boldsymbol{op}_{k,t_s}$ contains an array of binary variables $op_{k,t_s,i}$, which indicates whether the operation i is selected (1) or not (0). The stage with the minimum rate determines the rate of the line, $pr_k = \min_{t_s} spr_{k,t_s}$. We denote by $\boldsymbol{pr}(t)$, the processing rates of the line for all parts, as a measure of capacity at time t. By using s_{k,t_s} as an auxiliary variable representing the superfluous capacity, an optimization model is created to find the marginal capacity as

$$\min_{s_k,\boldsymbol{op}} - pr_k \tag{1}$$

$$\text{s.t.}(a)\ \mathbf{mp}_{k,i}^T \boldsymbol{op}_{k,t_s} \le \sum_{\tau_s=1}^{\tau_s=t_s-1} \mathbf{mc}_{k,i}^T \boldsymbol{op}_{k,\tau_s}, \forall t_s, \forall k, \forall i$$

$$(b) \sum_{t_s} \mathbf{M}_{ck} \boldsymbol{op}_{k,t_s} = 1, \forall k$$

$$(c)\ pr_k = spr_{k,t_s} - s_{k,t_s} = spr_{k,t_s'} - s_{k,t_s'}, \forall t_s', \forall k$$

$$(d) \sum_i op_{k,t_s,i} \le 1, op_{k,t_s,i} \in \{0,1\}, s_{k,t_s} \ge 0, \forall t_s, \forall k, \forall i,$$

where, $\mathbf{mp}_{k,i}$, $\mathbf{mc}_{k,i}$ are vectors that specify the precedence and type of operations for action i. 1 is a column vector with all entries as 1.. Constraint (a) ensures the operation actions occur in the required sequence; (b) ensures the completeness of the required operations; (c) ensures that the capacity of the production line is governed by the bottleneck stage, defined as the minimum processing rate across all stages; (d) bounds the feasible ranges of decision variables. Each RMS configuration is uniquely defined by the vectors $\boldsymbol{v}$ and $\boldsymbol{n}$ that record machine variants and numbers. This programming model in Eq. (1) allows us to link an RMS configuration to a maximum capacity, i.e., $pr_k = f_{pr}(k, \boldsymbol{v}, \boldsymbol{n})$ by determining the optimal set of operations that maximize the processing capacity for each target part type k.

3.3 Mining of Potential Configurations

By analyzing the marginal capacity associated with each potential system configuration, we proceed to identify the optimal configuration that maximizes the marginal capacity, subject to the available resource constraints. In this study, we investigate various RMS configurations to identify those that exhibit high capacity, utilizing the following nonlinear integer programming model.

$$\min_{\boldsymbol{v},\boldsymbol{n}} \sum_k w_k f_{pr}(k, \boldsymbol{v}, \boldsymbol{n}) \tag{2}$$

$$\text{s.t.}(a)\ f_{m_p}(\boldsymbol{v}, \boldsymbol{n}) \le s_{m_p}(t), \forall p$$

$$(b)\ 1 \le v_{t_s} \le N_v, 0 \le n_{t_s} \le N_n \text{and integer}, \forall t_s,$$

where, w_k represents the weight factor for part k, which can be dynamically tuned to prioritize the configuration designs in response to varying demands. f_{m_p} is a nonlinear function that calculates the required modules of type p for a given configuration. $s_{m_p}(t)$ denotes the available resources of type p on the shop floor at time t. Constraint (a) ensures that the required modules do not exceed the available stocks, and (b) bounds the feasible ranges of the decision variables. N_v is the number of machine variants. N_n is the maximum number of machines allowed per stage.

The optimization model in Eq. (1) which defines f_{pr} makes the cost function in Eq. (2) nonlinear and nonconvex. This complexity, combined with the nonlinearity introduced by constraint (a), makes it computationally intractable to efficiently find global optimal solutions. The primary objective of RMS configuration mining is not necessarily to find the global optimum but rather to identify a set of near-optimal configurations with high similarity, which can facilitate reconfiguration (Li et al. 2019). To achieve this, a genetic algorithm (GA) is used to generate potential solutions by the selection of a population of near-optimal configurations that share a high degree of similarity. However, increasing the number of machines enlarges the size of possible layouts, and GAs cannot efficiently output all solutions. Also, the module requirements for GA-generated solutions may not strictly match the available stocks, particularly for configurations generated in earlier populations. These earlier populations, however, create additional configurations that are potentially useful in scenarios with lower stock levels.

To enhance computational efficiency, we introduce a configuration library that stores all explored configurations by GA. This library includes a configuration index, machine layout, required modules, and capacity. Continuously updated throughout the planning process, this library functions as a dynamic knowledge base to support real-time decision-making. By enabling rapid retrieval of relevant configurations, it significantly improves the efficiency of the reconfiguration process and contributes to more effective and responsive planning.

3.4 Capacity Planning

Capacity planning involves assigning capacity at each time step during the planning horizon to minimize unsatisfied demands. The first priority is to select candidate capacities, which are the capacities in the configuration library that satisfy both the demand and module requirements. While these candidate capacities define the design space, it is still necessary to account for the daily processing time $\boldsymbol{pt}(t)$ which is influenced by the reconfiguration time $rt(t)$ spent each day. However, the nonlinearity of the problem and the high dimensionality of the decision variables make it computationally intractable to optimize all variables simultaneously. To address this, we decouple this problem by approximating the processing time and reconfiguration time as $\overline{\boldsymbol{pt}}(t)$ and $\overline{rt}(t)$ using heuristics. The detailed calculations for these times are provided in in Sects. 3.6 and 3.7. With this approximation, we formulate a binary integer programming model to minimize the expected total delayed demand during a planning horizon t_p as

$$\min_{\delta\boldsymbol{pr},\boldsymbol{s}_o,\boldsymbol{s}_b,n_d} \sum_{n_d}\left[\mathbf{c}_{\mathrm{o}}^T\boldsymbol{s}_{o,n_d} + \mathbf{c}_{\mathrm{b}}^T\boldsymbol{s}_{b,n_d}\right] + \sum_{t+1}^{t+t_p}\mathbf{c}_{\delta\mathbf{pr}}^T\delta\boldsymbol{pr}(\tau) \tag{3}$$

$$\text{s.t.}(a) \sum_{\tau=t+1}^{t_{d,n_d}}\left[\boldsymbol{pr}(\tau)\circ\overline{\boldsymbol{pt}}(\tau) - \boldsymbol{d}(\tau)\right] = \boldsymbol{s}_{o,n_d} - \boldsymbol{s}_{b,n_d}, \forall n_d$$

$$(b) \sum \boldsymbol{pr}(\tau) = 1, \forall\tau$$

$$(c)\ \boldsymbol{pr}(\tau+1) = \boldsymbol{pr}(\tau) + \mathbf{M}_{\delta\mathbf{pr}}\delta\boldsymbol{pr}(\tau), \forall\tau$$

$$(d)\ s_{o,n_d,j}, s_{b,n_d,j} \geq 0, \forall n_d, \forall j$$

$$(e)\ \delta pr_\alpha(\tau), pr_\alpha(\tau) \in \{0,1\}, \forall\tau, \forall\alpha,$$

where $\delta \boldsymbol{pr}(\tau)$ is a binary vector indicating the switch of capacity with cost $\mathbf{c}_{\delta \mathbf{pr}}$. $\mathbf{M}_{\delta \mathbf{pr}}$ records the changes of capacity under reconfiguration actions. n_d is the index of demand in the planning horizon. t_{d,n_d} is the due time of the n_dth demand. $\boldsymbol{d}(\tau)$ records all the received demands with due time as time step τ. $\boldsymbol{s}_{o,n_d}$, $\boldsymbol{s}_{b,n_d}$ are variables that represent overused and insufficient capacity of the n_dth demand, with costs $\mathbf{c}_o$ and $\mathbf{c}_b$. Constraint (a) states the balance between remaining demands and auxiliary variables. (b) comes from our assumption that RMS reconfiguration occurs only once for each time step τ. (c) states the balance of capacity at different times. (d) and (e) specify the type of decision variables. The proposed optimization model reveals a trade-off between the utility rate of capacity and the reconfiguration costs. By incorporating this trade-off into the objective function, the model is capable of adjusting the reconfiguration frequency based on the estimated cost of capacity mismatches.

3.5 Reconfiguration Action Planning

There are multiple configurations that achieve the same RMS capacity. Before deciding on the RMS configuration path, it is necessary to calculate the minimum time required to switch configurations. This minimum switching time was determined based on our previous research (Li and Epureanu 2017), and the optimization problem is formulated as a linear integer programming model:

$$\min_{\boldsymbol{oa}_{t_s},\ \boldsymbol{od}_{t_s}, \boldsymbol{oc}_{t_s}, \boldsymbol{ol}_{t_s}} \sum_{t_s} \boldsymbol{t}_a^T \boldsymbol{oa}_{t_s} + \boldsymbol{t}_d^T \boldsymbol{od}_{t_s} + \boldsymbol{t}_c^T \boldsymbol{oc}_{t_s} + \boldsymbol{t}_l^T \boldsymbol{ol}_{t_s} \tag{4}$$

$$\text{s.t.}(a)\ \boldsymbol{tc}'_{t_s} = \boldsymbol{tc}_{t_s} + \boldsymbol{oa}_{t_s} - \boldsymbol{od}_{t_s} + \mathbf{M}_c \boldsymbol{oc}_{t_s} + \mathbf{M}_l \boldsymbol{ol}_{t_s}, \forall t_s$$

$$(b)\ \sum_{t_s} \mathbf{m}_p \left(\boldsymbol{oa}_{t_s} - \boldsymbol{od}_{t_s} + \mathbf{M}_c \boldsymbol{oc}_{t_s} \right) \le s_{am_p}, \forall p$$

$$(c)\ \boldsymbol{oa}_{t_s}, \boldsymbol{od}_{t_s}, \boldsymbol{oc}_{t_s}, \boldsymbol{ol}_{t_s} \ge 0 \text{and integer}, \forall t_s,$$

where $\boldsymbol{tc}_{t_s}, \boldsymbol{tc}'_{t_s}$ record machine layouts before and after reconfiguration. $\boldsymbol{oa}_{t_s}, \boldsymbol{od}_{t_s}, \boldsymbol{oc}_{t_s}, \boldsymbol{ol}_{t_s}$ are decision variables for assembly, disassembly, reconfiguration and relocation. $\mathbf{M}_c$, $\mathbf{M}_l$ are matrices representing changes of machines due to reconfiguration and relocation actions. $\mathbf{m}_p$ records modules of type p required by each machine. $s_{am_p}(t)$ is the number of free modules of type p at time step t. Constraint (a) ensures that the reconfiguration is complete; (b) guarantees that the reconfiguration is achieved by the given free modules; (c) ensures all decisions are non-negative integers. Times required $\mathbf{rt}$ for reconfiguration actions are obtained for any pair of the candidates in the configuration library.

3.6 Configuration Path Planning

In configuration path planning, the goal is to select candidate configurations that effectively provide the target capacity $\boldsymbol{pr}(t)$. Given the evolving nature of the system configuration, we begin by modelling the changes in capacity as a dynamic system, which allows for the structured representation of the time-dependent adjustments in the system's configuration due to the reconfiguration activities and the associated changes in the functionality and capacity.

$$\boldsymbol{c}(t+1) = \boldsymbol{c}(t) + \mathbf{M}_{\delta \mathbf{c}} \delta \boldsymbol{c}(t), \tag{5}$$

$$\boldsymbol{cpr}(t+1) = \mathbf{M}_{\mathrm{pr}} \boldsymbol{c}(t) + \mathbf{M}_{\delta \mathbf{pr}} \delta \boldsymbol{c}(t), \tag{6}$$

where $\boldsymbol{c}(t)$ is a binary vector that indicates the selected candidate configuration with capacity $\boldsymbol{cpr}(t)$. $\mathbf{M}_{\delta \mathbf{c}}$ records how the configurations change under reconfiguration actions. $\mathbf{M}_{\mathbf{pr}}$ stores the capacity of all candidate configurations, which is estimated by the optimization model in Sect. 3.2. $\mathbf{M}_{\delta \mathbf{pr}}$ records the expected capacity loss from reconfiguration actions according to the reconfiguration time **rt** estimated in Sect. 3.5. Given all $\delta \boldsymbol{c}(t)$ over the planning horizon, we calculate the corresponding capacity by iteratively substituting Eq. (6). We formulate an integer programming model to minimize the total reconfiguration time as follows,

$$\min_{\delta \boldsymbol{c}} \sum_{\tau=t+1}^{t+t_p} \mathbf{rt}^T \delta \boldsymbol{c}(\tau) \tag{7}$$

$$\text{s.t.}(a)\ \boldsymbol{cpr}(\tau) = \boldsymbol{pr}(\tau), \forall \tau$$

$$(b) \sum \boldsymbol{c}(\tau) = 1, \forall \tau$$

$$(c)\ \delta c_\beta(\tau), c_\beta(\tau) \in \{0, 1\}, \forall \tau, \forall \beta.$$

Once the configuration path, represented by $\delta \boldsymbol{c}(\tau)$ is calculated, the exact reconfiguration time $rt(t)$ spent during each time step τ is determined also. That may differ from the estimated $\overline{rt}(t)$ used in Sect. 3.4. To improve the accuracy of these estimations, a neural network could be trained using historical records. The training dataset may include features such as inventory status, received demand profiles, current system configuration, and observed processing times. However, a detailed investigation of this machine learning-based estimation approach is beyond the scope of the current study.

3.7 Processing Time Planning

Due to potential inaccuracies in estimating reconfiguration times and unforeseen variations on the shop floor—such as fluctuations in individual machine processing speeds—the actual processing schedule must be dynamically adjusted at each time step τ, based on the previously determined reconfiguration path. To minimize unmet demand, the corresponding adjustment is formalized through the optimization model presented in Eq. (8).

$$\min_{pt_k} \sum_{\tau=t+1}^{t+t_p} \sum_k -\gamma^t pt_k(\tau) pr_k(\tau) \tag{8}$$

$$\text{s.t.}(a) \sum_{\tau=t+1}^{t_{d,n_d}} pt_k(\tau) pr_k(\tau) \leq \sum_{\tau=t+1}^{t_{d,n_d}} d_k(\tau), \forall n_d, \forall k$$

$$(b) \left[\sum_k pt_k(\tau) + rt(\tau) \right] \leq \bar{t}, \forall \tau$$

$$(c)\ pt_k(\tau) \geq 0, \forall \tau,$$

where the decision variable $pt_k(\tau)$ represents the production time for part k at time step τ. γ is a discount factor that incentivizes the system to finish jobs at earlier times. Constraint (a) limits the processed parts to be less than or equal to the demand. (b) bounds the total working time by a threshold $\bar{t}$. (c) ensures that the production times are non-negative. Thus, $\boldsymbol{f_u}$ used in Sect. 3.1 is formed as a function of the module stock $\boldsymbol{s}_m(t)$ and the demands in the planning horizon $\boldsymbol{D}(t)$, This function, though presented in a simplified format, encapsulates the models outlined in Sects. 3.2–3.7, namely,

$$\boldsymbol{f_u}(\boldsymbol{s}_m(t), \boldsymbol{D}(t)) = \sum_{n_d} \left[\sum_{\tau=t+1}^{t_{d,n_d}} [\boldsymbol{d}(\tau) - \boldsymbol{pt}(\tau) \circ \boldsymbol{pr}(\tau)] \right]. \tag{9}$$

4 Case Study

In this section, we implement the proposed model and simulate the real-time operation of a processing company to demonstrate how the capacity of RMS matches the demands in real-time and how the coordination between lines results in a better overall performance. We base our study on the RMS example presented by Youssef

Table 1 Modules and operation rates for different machine variants

		Machine variant								
		1	2	3	4	5	6	7	8	9
Module type	1	3	3	3	3	4	0	0	0	0
	2	1	2	3	4	1	0	0	0	0
	3	0	0	0	0	0	1	2	3	4
	4	1	1	1	1	1	0	0	0	0
	5	0	0	0	0	0	1	1	1	1
$\overline{or}_1$		11	22	33	44	12	2	4	5	7
$\overline{or}_2$		7	15	23	30	11	8	17	26	35

and ElMaraghy (2007), where two types of parts with multiple features require operation actions to process. We extend this study by assuming the availability of multiple RMSs for accomplishing the proposed processing tasks. The system comprises 2 types of machines and 9 variants in total, each with varying operation speeds. All the machines and their associated variants are constructed using 5 types of modules, as detailed in Table 1.

4.1 Real-Time Capacity Management

In this example, we consider a single RMS, which is initially provided with 50 modules of Type 1 and 20 modules of Types 2, 3, 4 and 5 to randomly arriving demands. The time interval between stochastically arriving demands is approximately 10 days, and the average completion time is one week. The magnitude of demands is generated based on Gaussian distributions, i.e., $d_k \sim N(5000{,}5000)$. The simulation spans over one year with a time step as 1 day. The average computation time required for daily reconfiguration decision-making is 0.42 s. Figure 3 shows the unfulfilled orders and their impact on the capacity of the RMS.

As shown in Fig. 3, the allocated capacity for Part B closely tracks its received demand throughout the year. This reflects the model's responsiveness and ability to dynamically reassign capacity to meet fluctuating requirements efficiently. No delayed demand was observed toward the end of the simulation period, indicating effective system responsiveness. In contrast, for Part A, there are periods—such as between Day 200 and Day 220—where capacity remains relatively high despite no immediate demand being received. This can be seen in the top subplot, where the blue line (allocated capacity) remains elevated while the orange bars (received demands) drop to zero. These instances suggest that the system deliberately maintains a buffer capacity for Part A. Such behavior arises from the model's capacity planning strategy in Sect. 3.4, which considers the high cost of frequent reconfiguration.

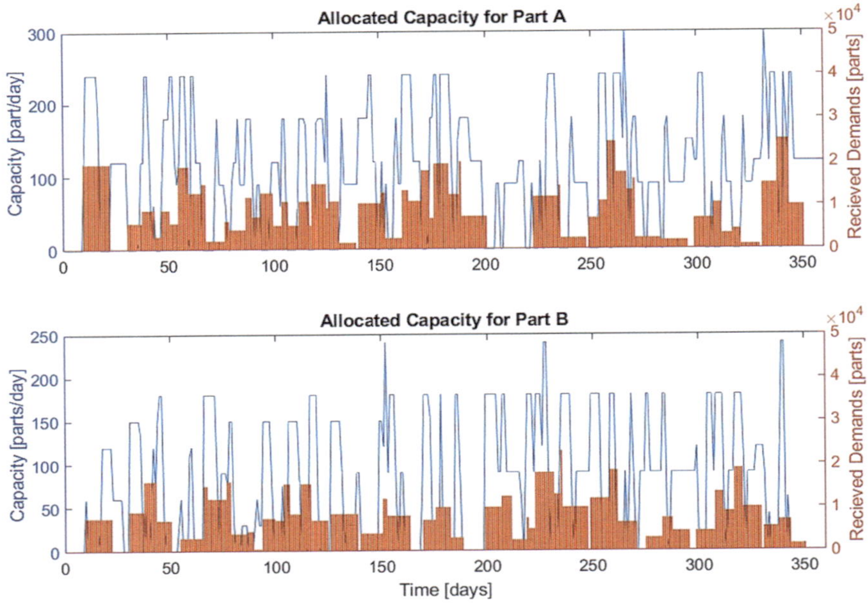

Fig. 3 Dynamic reconfiguration of a single RMS line to align system capacity with changing demands

Rather than constantly reallocating resources, the model balances the trade-off between reconfiguration cost and underutilized capacity. By retaining some capacity for Part A, the system ensures it can respond rapidly to future demand surges without incurring additional setup costs. This strategy underlines the flexibility and efficiency of the reconfigurable manufacturing system (RMS), enabling it to adjust operations while minimizing disruptions and maximizing long-term cost-effectiveness.

4.2 Coordination Among Lines (i.e., Among Multiple RMSs)

Moving to the second case, we consider the dynamic demand and resource reallocation among multiple RMSs. Here, we assume the demand settings are the same, and the firm allocates resources initially to the two processing lines as follows for Types 1–5: (a) line 1: 30, 10, 10, 10, 10, and (b) line 2: 20, 10, 10, 10, 10. Initially, the received demands are allocated evenly to each line. Because of the limited resources, one RMS may not accomplish the assigned task alone. Demand and resource reallocation are required to minimize the remaining demands via the bidding process, which increases the computational time required for decision-making to 2.79 s. We also considered the sudden increases of the processing demands.

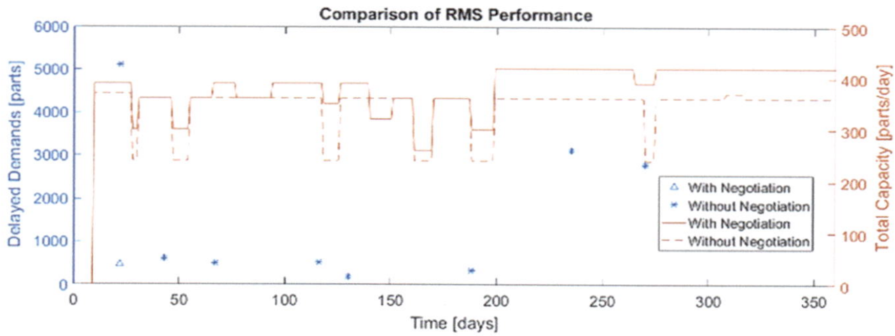

Fig. 4 Comparison of delayed demands and available capacity for two RMSs under scenarios with and without dynamic resource and demand reallocation

Figure 4 presents a comparison of RMS performance with and without negotiation under scenarios involving fluctuating and intensive demands. The RMS with negotiation demonstrates a significantly lower level of delayed demands throughout the observation period, as indicated by the scattered triangle markers on the left y-axis, compared to the consistently higher delayed demands in the non-negotiating system. This performance advantage stems from the negotiation-enabled system's ability to dynamically reallocate tasks and share manufacturing resources across system units. The right y-axis shows that the total capacity (solid line for negotiation, dashed line for no negotiation) is generally higher and more adaptive when negotiation is enabled. The negotiation mechanism allows for capacity ramp-ups during peak demand periods, minimizing delays without excessive idle capacity during low-demand periods. By facilitating resource exchange and task reassignment, the negotiation mechanism enhances the system's responsiveness, which is essential for managing the volatility of modern manufacturing environments.

Figure 5 provides a detailed view of the resource and demand relocation history driven by the negotiation-enabled bidding process. The stacked bars represent the negotiated demand allocations for Line 1 and Line 2 (left y-axis), while the lines track the total number of functional modules assigned to each line (right y-axis). In systems without negotiation, module allocations remain fixed (as shown by the constant red and red-dash-dot lines), limiting flexibility and capping the total capacity. In contrast, when negotiation is enabled, the system dynamically reallocates both demands and modules, as evidenced by the varying blue and blue-dash-dot lines. These real-time adjustments allow the system to temporarily ramp up capacity during peak demand periods—surpassing the fixed limit of 360 parts per hour and reaching up to 420 parts per hour—effectively mitigating unmet demands. This dynamic reallocation mechanism, powered by a bidding process, enables the RMS team to adapt its resource distribution in response to shifting demand patterns. The results demonstrate that, with dynamic reallocation, the RMS team is capable of adjusting its capabilities in real-time, enabling it to more effectively respond to fluctuating demands.

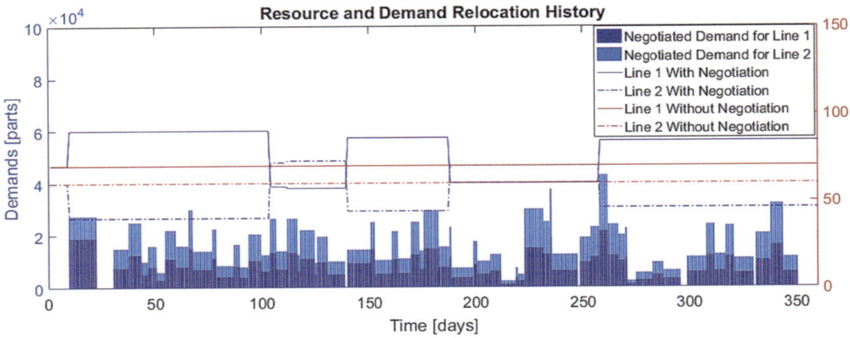

Fig. 5 Comparison of RMS capacity and unfulfilled demands w/o the bidding process

5 Conclusions

In this paper, we demonstrated the potential of autonomous RMSs for highly stochastic and rapidly changing demands. We formulated an integrated model for the decentralized management of multiple RMSs in real-time under limited resources. We formulated an integrated model for the autonomous, real-time coordination of autonomous RMS teams operating under constrained resources. Central to our approach is a novel bidding mechanism that enables dynamic relocation of both resources and demands across multiple production lines, thereby enhancing overall system responsiveness and utilization. The results underscore the effectiveness of dynamically networked RMSs in optimizing performance, maintaining high readiness levels, and maximizing utility in the face of fluctuating operational requirements.

Future work will extend this foundation by incorporating generative artificial intelligence to support autonomous system configuration and adaptive path planning. By enabling the system to generate and evaluate a broader set of reconfiguration strategies in real time, this approach aims to significantly enhance the scalability of RMS operations—especially under complex and unpredictable demand scenarios, as well as in response to human-in-the-loop interactions (Li et al. 2024a, b).

References

Asl M, Ulsoy G (2003) Stochastic optimal capacity management in reconfigurable manufacturing systems. CIRP Ann Manuf Technol 52(1):371–374

Azab A, ElMaraghy HA (2007) Mathematical modelling for reconfigurable process planning. CIRP Ann Manuf Technol 56(1):467–472

Bensmaine A, Dahane M, Benyoucef L (2014) A new heuristic for integrated process planning and scheduling in reconfigurable manufacturing systems. Int J Prod Res 52(12):3583–3594

Ceryan O, Koren Y (2009) Manufacturing capacity planning strategies. CIRP Ann Manuf Technol 58(1):403–406

Epureanu BI, Li X, Nassehi A, Koren Y (2021) An agile production network enabled by reconfigurable manufacturing systems. CIRP Ann 70(1):403–406

Gu X, Koren Y (2022) Mass-individualisation–the twenty first century manufacturing paradigm. Int J Prod Res 60(24):7572–7587

Inoue S, Urata A, Kodama T, Huwer T, Maruyama Y, Fujita S, Shinno H, Yoshioka H (2021) High-precision mobile robotic manipulator for reconfigurable manufacturing systems. Int J Autom Technol 15(5):651–660

Kim YH, Ye W, Kumar R, Bail F, Dvorak J, Tan Y, May MC, Chang Q, Athinarayanan R, Lanza G (2024) Unlocking the potential of remanufacturing through machine learning and data-driven models—a survey. Sched Theory Algorithms Sustain Manuf 17:28

Koren Y, Shpitalni M (2010) Design of reconfigurable manufacturing systems. J Manuf Syst 29(4):130–141

Koren Y, Heisel U, Jovane F, Moriwaki T, Pritschow G, Ulsoy G, Van Brussel H (1999) Reconfigurable manufacturing systems. CIRP Ann Manuf Technol 48(2):527–540

Koren Y, Gu X, Guo W (2017) Reconfigurable manufacturing systems: principles, design, and future trends. Front Mech Eng

Li X, Bayrak AE, Epureanu BI, Koren Y (2018) Real-time teaming of multiple reconfigurable manufacturing systems. CIRP Ann 67(1):437–440

Li X, Nassehi A, Epureanu BI (2019) Degradation-aware decision making in reconfigurable manufacturing systems. CIRP Ann 68(1):431–434

Li X, Athinarayanan R, Wang B, Yuan W, Zhou Q, Jun M, Bravo J, Gao RX, Wang L, Koren Y (2024a) Smart reconfigurable manufacturing: literature analysis. Procedia CIRP 121:43–48

Li X, Epureanu BI (2017) Robustness and adaptability analysis of future military modular fleet operation system. In: ASME 2017 dynamic systems and control conference. American Society of Mechanical Engineers

Li X, Nassehi A, Wang B, Hu SJ, Epureanu BI (2023) Human-centric manufacturing for human-system coevolution in Industry 5.0. CIRP Ann 72(1):393–396

Li X, Tao F, Ye W, Nassehi A, Sutherland JW (2024) Generative manufacturing systems using diffusion models and ChatGPT. arXiv:2405.00958

Li X, Nassehi A, Hu SJ, Joung BG, Gao RX (2025) A large manufacturing decision model for human-centric decision-making. CIRP Ann

Terzic I, Zoitl A, Rooker M, Strasser T, Vrba P, Mařík V (2009) Usability of multi-agent based control systems in industrial automation. In: Holonic and multi-agent systems for manufacturing: 4th international conference on industrial applications of Holonic and multi-agent systems, vol 4, pp 25–36. HoloMAS 2009, Linz, Austria, August 31-September 2, 2009. (Proceedings)

Wang W, Koren Y (2012) Scalability planning for reconfigurable manufacturing systems. J Manuf Syst 31(2):83–91

Youssef AM, ElMaraghy HA (2007) Optimal configuration selection for reconfigurable manufacturing systems. Int J Flex Manuf Syst 19(2):67–106

A Multi-Agent Deep Reinforcement Learning for Flexible Job Shop Scheduling Problem with Transportation

QiXin Wang, Peng Guo, ZhiLin Ma, and XiangHeng Meng

Abstract In this paper, we propose a multi-agent Deep Reinforcement Learning (DRL) framework to address the Flexible Job Shop Scheduling Problem with Transportation (FJSPT). We decompose the FJSPT into three sub-problems: machine selection, job sequencing, and Automated Guided Vehicle (AGV) scheduling. To model the shop floor layout and machine state features, we employ a graph attention neural network. A policy gradient-based algorithm is used to train the machine selection sub-policy. For job sequencing, we design a transformer-based structure to embed job state features and facilitate action selection, with a Proximal Policy Optimization (PPO) algorithm used to train the job sorting sub-policy. Finally, we develop effective scheduling rules for AGV selection. Experimental results demonstrate that, after individual training, the two agents in our framework outperform handcrafted heuristic dispatching rules in terms of solution quality. Upon completing centralized training, the framework delivers solutions that are qualitatively comparable to state-of-the-art heuristic algorithms while maintaining a short running time. Additionally, the model trained on randomly generated instances performs better on both large-scale and benchmark instances, highlighting the framework's strong generalization capabilities.

1 Introduction

Flexible manufacturing, characterized by process flexibility, machine flexibility, and operational flexibility, facilitates the effective coordination and utilization of diverse production resources. It is widely employed in semiconductor, automotive, textile, and other manufacturing sectors, and has emerged as a prominent trend in modern manufacturing. In flexible manufacturing systems, the Flexible Job Shop

Q. Wang · P. Guo (✉) · Z. Ma · X. Meng
School of Mechanical Engineering, Southwest Jiaotong University, Chengdu, China
e-mail: pengguo318@swjtu.edu.cn

S. Huang et al. (eds.), *Design and Operation of Smart Reconfigurable Manufacturing Systems in Industry 4.0/5.0*, Springer Series in Advanced Manufacturing,
https://doi.org/10.1007/978-3-032-00284-6_7

Scheduling Problem (FJSP) is a typical job shop scheduling problem and a well-established combinatorial optimization problem in operations research, which has garnered significant attention and research over the years.

In a typical FJSP, each operation of a job must be assigned to one of several compatible machines, and the processing sequence of machines must be optimally arranged to minimize objectives such as completion time and tardiness. However, most theoretical studies simplify the problem by overlooking real-world factors, such as transportation time and setup time, resulting in a gap between the models and actual production processes.

In recent years, driven by advancements in intelligent manufacturing technologies such as information technology and robotics, many manufacturing enterprises have begun using Automated Guided Vehicles (AGVs) in flexible job shops to automate the transportation of parts and materials, thereby enhancing production automation and efficiency (Lacomme et al. 2013). At the same time, due to the limited number of AGVs and the high demand for transportation, efficiently scheduling AGV transportation processes can significantly improve both transportation and overall production efficiency. Therefore, incorporating transportation time into the Flexible Job Shop Scheduling Problem with Transportation (FJSPT) better aligns with actual production processes and holds significant practical research value. This paper primarily investigates the FJSPT, where the number of AGVs is constrained, necessitating the simultaneous consideration of job machine allocation, processing sequence, and AGV transportation order. Compared to the FJSP, the complexity of this problem is significantly increased.

Over the past decades, numerous studies have been conducted on job shop scheduling, a combinatorial optimization problem. Among these, the most common scheduling methods are scheduling rules and heuristic scheduling algorithms. Scheduling rules are straightforward methods, designed based on domain knowledge and scheduling experience, that aim to provide reasonable solutions. Common scheduling rules include Shortest Processing Time (SPT), First-In-First-Out (FIFO), and Earliest Delivery Date (EDD), among others. These rules can quickly generate scheduling solutions, making them widely used in real-world production. However, scheduling schemes based on these rules often lack optimization and may not provide stable performance across different problem instances, leaving considerable room for improvement (Luo 2020). In addition, numerous heuristic algorithms, such as the Genetic Algorithm (GA), Particle Swarm Optimization (PSO), and Adaptive Large Neighborhood Search (ALNS), have been applied to solve shop floor scheduling optimization problems. Typically, these heuristic algorithms can obtain near-optimal scheduling solutions through multiple search iterations. However, their ability to produce high-quality results comes with significantly high time complexity, particularly for large-scale real-world scheduling problems. Due to the vast solution space, it is challenging to find optimal solutions within a reasonable timeframe (Gao et al. 2019).

In recent years, with the continuous development of Deep Learning (DL) and Reinforcement Learning (RL), numerous studies have applied these techniques to solve combinatorial optimization problems, such as the Traveling Salesman Problem

(TSP), Vehicle Routing Problem (VRP), and their various extensions (Bengio et al. 2021). In RL, the Markov Decision Process (MDP) is the most commonly used mathematical framework for modeling the interaction between an intelligent agent and its environment. In an MDP, the agent selects an action based on the current state, receives a reward, updates the environment's state, and uses the training algorithm to learn the optimal strategy. DL can effectively extract complex state features and is widely used in fields such as machine vision and natural language processing. By combining reinforcement learning with deep learning, and designing effective neural network models to extract these complex features in combinatorial optimization problems, Deep Reinforcement Learning (DRL) offers a novel approach to solving shop floor scheduling problems and their extensions (Yan et al. 2021).

Most of the current DRL-based scheduling optimization algorithms focus on the FJSP, and these studies mainly involve two types of methods: end-to-end and rule-based approaches. In the end-to-end method, the process to be executed and the processing equipment are directly selected at each time step. In contrast, the rule-based method selects from multiple rules at each time step and then determines the process and equipment based on the chosen rule. The effectiveness of the rule-based approach largely depends on the quality of the designed scheduling rules, requiring substantial domain knowledge and experience. In contrast, the end-to-end approach better leverages the advantages of DRL (Lee and Lee 2022; Lei et al. 2022a, b; Nguyen et al. 2021). However, for FJSPT, which involves three decisions—process selection, machine selection, and AGV selection—its action space is vast, especially in the case of large problem sizes, making it difficult for general deep reinforcement learning methods to ensure training stability. In addition, most current scheduling problems based on deep reinforcement learning select actions at each time step until the scheduling is completed. However, in FJSPT, this approach may lead to inefficient use of AGV resources, resulting in excessive idle time for the AGVs.

To address the aforementioned problem, we propose a distributed multi-agent deep reinforcement learning framework for solving FSJPT. In this framework, two main agents are designed: Machine Selection Agent (MSA) and Task Selection Agent (TSA) Firstly, MSA selects the machining machines for all the operations to be processed, and then TSA sorts the tasks on each machine. For AGV selection, we design efficient scheduling rules to reduce the action space. Overall, this paper proposes an effective multi-agent deep reinforcement learning framework to learn two strategies for solving FJSPT. The main contributions are as follows:

Proposes a distributed deep reinforcement learning framework with multiple agents, where two agents handle the decisions of machine selection and processing sequencing, respectively. This reduces the action space during decision-making and contributes to training stability, while also enabling the solution of larger-scale problems.

For the transportation distance between machines in FJSPT, an encoder based on a graph attention neural network is used in MSA to embed the features of each machine node while effectively aggregating the information from neighboring nodes.

During the model training phase, to effectively train the two sub-strategies, we propose a decentralized-concentrated training approach. This approach first trains the two sub-strategies in a distributed manner to verify their validity, and then trains them centrally, verifying their performance on standard datasets and large-scale problem instances.

2 Literature Review

2.1 The FJSPT

The FJSPT was first addressed by Deroussi and Norre (2010), who proposed an iterated local search algorithm that includes local search, perturbation, and an acceptance criterion. Additionally, they introduced a set of benchmark instances that have been widely used in subsequent research. Kumar et al. (2011) combined differential evolutionary algorithms with machine selection heuristics and vehicle allocation heuristics effectively utilizes machine and vehicle resources to shorten the production cycle. Nouri et al. (2016) proposed a hybrid meta-heuristic algorithm based on the clustered holographic multi-agent model, which first explores the search space globally using a Neighborhood-based Genetic Algorithm (NGA), and then further completes the exploration using tabu search technique. Huang and Yang (2019) combined a genetic algorithm with a simulated annealing algorithm to solve the multi-objective flexible job shop scheduling problem considering transportation time, and experimentally demonstrate that the proposed hybrid genetic algorithm outperforms the original GA. Homayouni et al. (2023) proposed an operation-based multi-start biased random key genetic algorithm (BRKGA) to solve FJSPT, which was tested on several standard problem instances to find high-quality solutions in a short period of time. Dai et al. (2018, 2019) developed a multi-objective optimization model for FJSPT to minimize energy consumption and completion time, and they designed an improved genetic algorithm to solve it. Homayouni and Fonte (2021) proposed a mixed-integer linear programming model for FJSPT to obtain optimal solutions for small-scale problem instances. Additionally, they introduced a local search algorithm to solve larger-scale problem instances. Ren et al. (2022) proposed a particle swarm optimization algorithm incorporating genetic operators to solve the dynamic FJSPT and experimentally compared it with commonly used scheduling rules and heuristic algorithms to verify its effectiveness. Chaudhry et al. (2022) proposed a solution based on Microsoft Excel spreadsheets and a genetic algorithm to optimize scheduling. Berterottiere et al. (2023) proposed a meta-heuristic algorithm using a neighborhood function capable of exploring a large number of moves, validated the algorithm, and a new benchmark instance was proposed. The above studies have employed traditional methods such as mathematical programming, heuristics, and metaheuristics, which are not suitable for large-scale or real-time instances due to the complexity of the FJSPT and the challenges in ensuring the efficiency of solutions using these methods. Recently, with the advancement of AI technology, several studies have applied DRL to address FJSPT-related problems. Du et al. (2022) proposed a multi-objective FJSP model that incorporates crane transportation and installation time, and introduced a Deep Q-Network (DQN) model to select appropriate scheduling rules for different scheduling scenarios. Li et al. (2022) proposed a hybrid deep Q network (HDQN) combining DQN with three extensions, and their experimental results showed the superiority and versatility of HDQ compared to current optimization-based methods. Zhang et al. (2023) addressed the distributed flexible job-shop scheduling problem with crane transportation (DFJSPC) by constructing a mixed-integer linear programming model and proposed a Q-learning-based hyper-heuristic evolutionary algorithm (QHHEA) to solve the problem, which was tested and compared on a dataset containing 36 instances. All of the above DRL-based approaches use DQN-based models to select scheduling rules or combine them with heuristic algorithms, making efficient scheduling rules and good initial solutions crucial. Achieving this typically requires strong domain knowledge and a thorough analysis of the specific problem.

2.2 DRL for Scheduling Problem

The scheduling problem, as a typical combinatorial optimization problem, has long attracted widespread attention. Numerous studies have employed traditional methods, such as scheduling rules, exact algorithms, and approximation methods, to address this issue. With the application of DRL to combinatorial optimization problems like the Traveling Salesman Problem (TSP), Vehicle Routing Problem (VRP), and graph optimization, along with its demonstrated advantages in solving efficiency and generalization performance, an increasing number of studies are now using DRL to tackle scheduling problems. Hu et al. (2020) used DQN to solve the dynamic scheduling problem of flexible manufacturing systems (FMSs) and achieved some advantages over traditional methods in terms of computational efficiency and adaptability. Wang (2020) established a dynamic scheduling system model based on multi-agent technology for the shop floor scheduling problem and proposed a weighted Q-learning algorithm that incorporates clustering and dynamic search to select scheduling operations. The algorithm was validated in various scheduling environments to verify its effectiveness. Pan et al. (2023) proposed an efficient DRL-based optimization algorithm for solving the permutation flow-shop scheduling problem (PFSP), which achieves better solutions in similar computation time compared to existing heuristic algorithms. Liu et al. (2022) proposed a hierarchical distributed architecture for solving the dynamic flexible job shop scheduling problem, using a Double Deep Q-Network algorithm to train the scheduling agents. The algorithm demonstrates strong adaptive capabilities. Wang et al. (2022) proposed a dynamic multi-objective flexible job shop scheduling problem to simulate the real production environment, along with a DRL-based multi-objective scheduling algorithm. The algorithm was validated on 27 instances, demonstrating superiority and stability compared to other algorithms. Chen et al. (2023) introduced a new deep reinforcement learning framework to solve the classical job shop scheduling problem (JSSP) based on an attention mechanism and analytic graph embedding, using sequence-to-sequence modeling for JSSP within the framework. Song et al. (2023) proposed a heterogeneous graph neural network-based architecture to capture the complex relationship between operations and machines during scheduling and designed a DRL-based end-to-end algorithm to solve the FJSP. The algorithm demonstrated superior solution efficiency and generalization performance compared to traditional prioritized scheduling rules in the experiments. Lei et al. (2024) developed a novel end-to-end hierarchical reinforcement learning framework for solving large-scale dynamic flexible job shop scheduling (DFJSP) problems in near real-time. In this framework, a high-level agent is designed to automatically decompose the DFJSP into sub-problems, i.e., static FJSPs of different sizes, and two low-level agents are constructed to solve these sub-problems. From the above literature summary, it is evident that DRL-based algorithms have achieved some success in the field of production scheduling. However, most studies focus on the FJSP problem and rarely consider the transportation aspect in the scheduling process. Additionally, most of the studies use DRL to select predefined scheduling rules. In contrast, this paper designs an end-to-end algorithmic framework to efficiently solve the more complex state characteristics and action space of the FJSPT.

3 Preliminary

3.1 Problem Description

To facilitate the following description of the FJSPT, the symbols used are first given, as shown in Table 1.

In this paper FJSPT can be described as follows: there are n jobs to be processed in the shop $J = \{J_1, \ldots, J_n\}$, m available processing machines $M = \{M_1, \ldots, M_m\}$, and u AGV $V = \{V_1, \ldots, V_u\}$ for transporting the jobs. Each job J_i comprises n_i successive processes $O_i = \{O_{i1}, \ldots, O_{in_i}\}$ with priority constraints, one or more available processing machines for each operation O_{ij}, and the processing time of the process on machine k is denoted by t_{ijk}. When successive processes of a job are not processed on the same machine, AGVs are required to transfer the job, and the processing time of the process on machine k is denoted by t_{ijk}. When successive operations of a job are not processed on the same machine, AGVs are required to transfer the job, and the transport time of the AGV from machine k to machine $k\prime$ is denoted by $t_{Kk\prime}$. In addition, all the jobs are initially located in the load/unload (LU) area, and the jobs need to be transported from the LU to the corresponding machines when they start the process. Figure 1 is a schematic diagram of an FJSPT with 4 machines.

In FJSPT, it is necessary to assign all operations of the jobs to the available machines, arrange the processing sequence to determine the start and end times of each operation, select the transporting AGVs for jobs that need to be transferred, and

Table 1 Symbols and descriptions

Notation	Description
J	Job set, $J = \{J_1, \ldots, J_n\}$
M	Machine set, $M = \{M_1, \ldots, M_m\}$
V	AGV set $V = \{V_1, \ldots, V_u\}$
O_i	Operation set for job J_i, $O_i = \left\{O_{i1}, \ldots, O_{in_i}\right\}$
W	Operation set on machine k, $W_k = \left\{O_{ij}, \ldots, O_{i'j'}\right\}$
n	Number of jobs
m	Number of machines
u	Number of AGVS
n_i	Number of operations for J_i
i	Job index, $i = 1, \ldots, n$
j	Operation index for $J_i, j = 1, \ldots, n_i$
l	Machine index,$k = 1, \ldots, m$
v	AGV index,$v = 1, \ldots, u$
t_{ijk}	Processing time of operation O_{ij} on machine k
$t_{Kk\prime}$	Transport time from machine k to machine $k\prime$

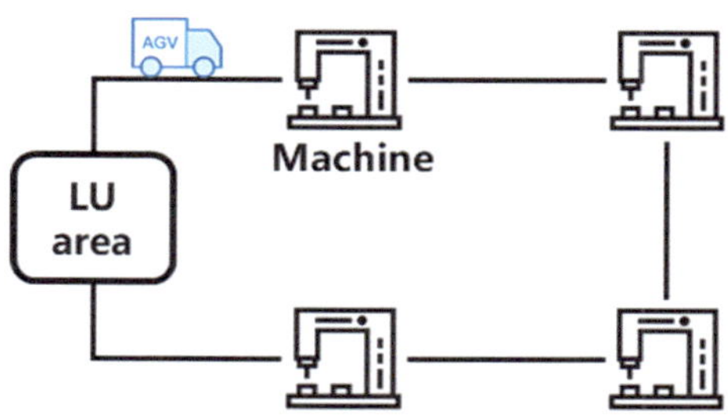

Fig. 1 A diagram of FJSPT

determine the sequence in which the AGVs transport the jobs, all with the goal of minimizing the completion time.

In addition, the FJSPT in this study fulfills the following assumptions: (1) All the jobs are independent and available at time zero, (2) All the machines and AGVs are available at time zero, (3) Only one job can be processed or transported by each machine and AGV at the same moment, (4) Jobs in processing cannot be interrupted, (5) All AGVs have the same transportation speed, (6) job loading, unloading and setup time is included in the processing time.

3.2 Reinforcement Learning and Graph Attention Networks

Reinforcement learning is a key branch of machine learning that enables agents to learn by interacting with their environment in order to achieve specific goals. In reinforcement learning, an agent explores the environment by taking different actions and learns which actions yield better outcomes based on feedback from the environment. Generally, reinforcement learning consists of five components: the agent, environment, state, action, and reward. At each time step, the agent takes an action based on the state information provided by the environment, receives a corresponding reward, and updates the environment and state based on the chosen action. Through repeated interactions, the agent refines its decision-making process. The ultimate goal of reinforcement learning is to maximize the cumulative reward, and various optimization strategies can be employed as long as the reward is designed to align with the desired objective.

The Graph Attention Network (GAT) is a neural network model designed for processing graph-structured data. It performs aggregation operations on neighboring nodes using the attention mechanism, allowing for adaptive weight allocation to different neighbors, which significantly enhances the expressive power of the graph neural network model (Veličković et al. 2018). In a graph with m nodes, the importance between nodes x, y is calculated using the attention coefficient as follows

$$\alpha_{xy} = \frac{\exp(LeakyReLU\,(a^T[Wh_x||Wh_y]))}{\sum_{z=1}^{m}\exp(LeakyReLU\,(a^T[Wh_x||Wh_z]))} \tag{1}$$

where LeakyReLU is the activation function, a^T and W are the weight vectors and matrices, and || denotes the matrix splicing operation. For node h_x, its updated feature output is:

$$h'_x = \sigma\left(\sum_{y=1}^{m}\alpha_{xy}W_1h_y\right) \tag{2}$$

where σ is the activation function and W_1 is the weight matrix.

4 Proposed Method

Based on the research outlined above, we propose a multi-agent deep reinforcement learning framework, which consists of a Machine Selection Agent (MSA) and a Task Selection Agent (TSA). These two agents collaborate to solve the FJSPT. Specifically, the MSA assigns each operation of all jobs to a machine, considering factors such as processing time on each machine, transportation distances between machines, and machine load—essentially determining the processing path for each job. Meanwhile, the TSA determines the processing order of operations on each machine. By decomposing the complex FJSPT into three sub-problems, this approach facilitates a more efficient solution. Next, we formulate the scheduling process for the two agents as an MDP detailing their states, actions, and rewards. We then introduce the encoding and decoding strategies for both agents, along with the training process.

4.1 Machine Selection Agent (MSA)

At time step t, the MSA selects a machine for an unassigned operation based on the current state of the environment. After executing the action, the environment state is updated, and the corresponding reward is received. This process is repeated until all operations are assigned to available machines, which can be formulated as the following MDP.

(1) The MDP definition for MSA

State: for MSA, at each time step its state is represented as s_t^* including two parts: machine feature $s_t^{*,m}$ and process feature $s_t^{*,o}$. Where $s_t^{*,m}$ is represented by the machine layout $G_T(V, E)$, where V denotes the machine node in the graph, the node feature is the current machine load, and E denotes the distance between the machines. $s_t^{*,o}$ is a two-dimensional feature vector for each operation with respect to the machine, including both the processing time and transportation time.

Action: the action $a_t^* = \left(O_{ij}, M_l\right)$, denotes the assignment of the operation O_{ij} to be processed on the machineM_l, where at each time step t, we sequentially select the operation, so that in the actiona_t^*, the agent only needs to select the compatible machineM_l.

Reward: The MSA evaluates the suitability of a machine for assigning an operation. In FJSPT, the optimization objective is to minimize the maximum completion time. Therefore, for the machine assignment action in MSA, we aim to balance the machining time across all machines as much as possible. Additionally, due to the limited number of AGVs, the frequency and distance of job transportation must also be considered. The forward r_t^* for time step t is calculated as follows:

$$r_t^* = \max\left\{M_{t,l}^*\right\} - \max\left\{M_{t+1,l}^*\right\} \tag{3}$$

where $M_{t,k}^*$ denotes at time step t, the sum of processing time for all processes on machine l and transportation time to machine l.

(2) The encoder-decoder structure

To effectively handle feature inputs with a graph structure, we designed an encoder-decoder structure based on GAT and the attention mechanism. Figure 2 illustrates the encoding–decoding process, and the following sections will provide a detailed description of the components of the encoder-decoder.

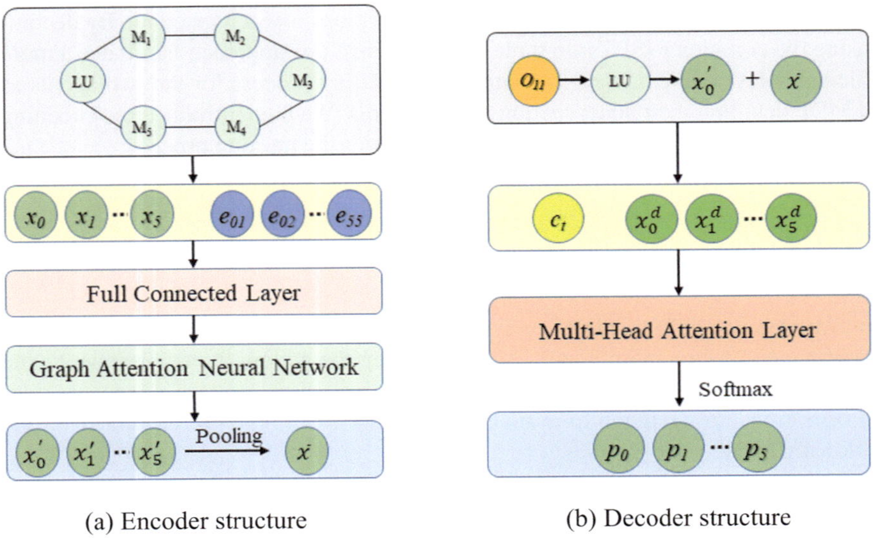

Fig. 2 The encoder-decoder structure of MSA

The machine features in the scheduling process are represented by a machine layout graph. To effectively extract these features, a Graph Attention Network (GAT) is used for embedding, while machine distance features are incorporated by adding edge information into the GAT. Firstly, the node feature x_l as well as the edge feature $e_{ll'}$ are embedded through the fully connected layer, and after the embedded features are input into GAT, Eqs. (4) and (5) describe the embedding computation of nodes and edges.

$$x_l = BN(A_0 x_l + b_0), l \in \{1, \dots, m\} \tag{4}$$

$$e_{ll'} = BN(A_1 e_{ll'} + b_1), l, l' \in \{1, \dots, m\} \tag{5}$$

where A_0 and A_1, b_0 and b_1 are the weight matrix and vector respectively, $BN(\cdot)$ denotes the batch normalization (batch normalization) operation, and the special x_0 is the feature embedding of the initial node LU. Then the feature embeddings of nodes and edges are inputted into GAT, and the attention coefficients between nodes are first calculated and then aggregated to obtain each node embedding x_l', and finally the graph embedding vector $\overline{x}$ is obtained, Eqs. (6)–(8) show the node feature aggregation as well as the calculation of $\overline{x}$.

$$\alpha_{ll'} = \frac{\exp\left(LeakyReLU\left(a^T[W_0 x_l || W_0 x_{l'} || W_0 e_{ll'}]\right)\right)}{\sum_{z=1}^{m} \exp\left(LeakyReLU\left(a^T[W_0 x_l || W_0 x_z || W_0 e_{lz}]\right)\right)} \tag{6}$$

$$x_l' = \sigma\left(\sum_{l'=1}^{m} \alpha_{ll'} W_1 x_{l'}\right) \tag{7}$$

$$\overline{x} = \frac{1}{m}\sum_{l=1}^{m} x_x' \tag{8}$$

where LeakyReLU and σ are activation functions, a^T is weight vectors, W_0 and W_1 are weight matrices, and || denotes the splicing operation of the matrices.

After obtaining the output of the encoder at each time step t, it is fed into the decoder, we design a decoder structure based on the multi-head attention mechanism and refer to the decoding of pointer networks, and select a machine as the output of decoding at each step according to the *softmax* output probability distribution, and we describe the decoding process in detail next.

Firstly, the context vector c_t is computed based on the graph embedding vector $\overline{x^*}$ of the machine output by the encoder and the encoding $x_l^{*,\prime}$ of the machine where the process O_{ij} is to be scheduled at time step t. The dynamic encoding of each machine at time t is computed based on the machining time of the process O_{ij} at each machine x_l^d:

$$c_t = \overline{x} + x_{l'} \tag{9}$$

$$x_l^d = x_{l'} + W_d D_{ijl} \tag{10}$$

where W_d is the weight matrix and D_{ijl} is the two-dimensional feature vector of process O_{ij} corresponding to machine l, including processing time and transportation time. Then the query vector q, the key vector k_l and value vector v_l are computed using c_t, x_l^d:

$$q = W^Q c_t, k_l = W^K x_l^d, v_l = W^V x_l^d, l \in \{1, \ldots, m\} \tag{11}$$

where W^Q, W^K, W^V are learnable weights. Calculate the attention factor u_l based on the obtained query vector q, the key vector k_l:

$$u_l = \begin{cases} \frac{q^T k_l}{\sqrt{d_q}}, & If machine l can process O_{ij} \\ -\infty, & otherwise \end{cases} \tag{12}$$

where q^T is the transpose of q, d_q is the first dimension of W^Q, and to avoid selecting a machine that can not process O_{ij} we set its u_l to $-\infty$, and according to Eqs. (12) and (13), calculate the H-attention coefficients and sum them to get the multi-attention coefficients $u_{l'}$, and finally get the probability of the selection of each machine by the softmax function

$$u_l' = \sum_{h=1}^{H} u_l^h \tag{13}$$

$$p_l = softmax\big(C \cdot \tan h\big(u_l'\big)\big), l \in \{1, \ldots, m\} \tag{14}$$

where tan$h()$ is the activation function, we used a hyper-parameter C to clip the result of the activation function. A random adoption strategy is used to select the machine during training to improve the exploration performance of the agent, while a greedy adoption strategy is used to select the machine during the testing phase to ensure the stability of the solution results.

(3) Training methods

In the training method, the reinforcement learning algorithm based on policy gradient is adopted, and the loss function is defined as $Loss^*_{(\theta|S)} = \mathbb{E}_{\pi \sim p_\theta(\pi^*|s^*)}[R(\pi^*|s^*)]$, meanwhile, in order to accelerate the convergence of the algorithm and to improve the stability, Rollout's baseline method is adopted at the same time to improve the training efficiency, and the loss function of the gradient is calculated as follows:

$$\nabla_\theta Loss^*_{(\theta|S)} = \mathbb{E}_{\pi \sim p_\theta(\pi^*|s^*)}\big[\big(R\big(\pi^*|s^*\big) - b\big)\nabla_\theta logp_\theta\left(\pi^*|s^*\right)\big]$$

where b represents the result obtained from the critic network. The critic and actor share the same architecture and initialization parameters. After updating the actor's parameters in each epoch, a t-test is conducted to compare it with the critic. If the actor's policy network shows a significant improvement in the test cases, the actor's parameters are copied to the critic to complete the update.

4.2 Task Selection Agent

At time step t, the TSA selects a job to be processed based on the current state of the environment. After executing the action, the environment state is updated, and the corresponding reward is received. This process is repeated until all jobs have been processed. The MDP for this process is as follows:

(1) The MDP definition for TSA

State: for TSA at each time step t, the agent makes a decision to select the machining order of the job based on the current state, where both the job and the state of the machine in which the job is located are considered up to s_t^∇, s_t^∇ consists of the feature vector h_i for each job, and h_i contains 5 elements:

Minimum time that job i has to wait for agv: $T_{i,t}^w$

Current process end time of job i: $T_{i,t}^f$

Time for job i to reach the next processing machine: $T_{i,t}^a$

Number of processes remaining on job i: $R^{o}_{i,t}$

Remaining machining time for job i: $R^{p}_{i,t}$

Action: at each time step t, the action a^{∇}_{t} is to select one of the unfinished jobs for machining and then update the environment state.

Reward: in order to minimize the completion time, the corresponding reward is designed as the value of the change in the maximum completion time between time steps t and t + 1:

$$r^{\nabla}_{t} = C_{max}\left(s^{\nabla}_{t}\right) - C_{max}\left(s^{\nabla}_{t+1}\right) \tag{15}$$

(2) The encoder-decoder structure

In the job scheduling process, the state feature of each job is represented by a 5-dimensional feature vector s^{∇}_{t}. To efficiently extract feature information, we design an encoder-decoder structure based on the Transformer model, a deep learning architecture introduced by Vaswani et al. in (2017). The Transformer excels in processing sequential data, capturing long-term dependencies, and enabling efficient parallel computation, and it has achieved significant success in the field of natural language processing (NLP). For combinatorial optimization problems, such as TSP, VRP, and other path-sequence problems, the task can be viewed as a sequence. Similarly, in this section, the processing order of jobs can also be framed as a sequential decision-making problem, making it feasible to adopt a Transformer-based structure. The encoder structure is shown in Fig. 3.

First, the feature vector h_i of each job, is input to a fully connected layer neural network with a learnable weight matrix and vectors, and normalized to obtain the initial node embedding h^{0}_{i}:

$$h^{0}_{i} = BN(A_2 h_i + b_2) \tag{16}$$

Next, h^{0}_{i} is computed through the attention layer to obtain the node embedding h^{1}_{i} in the next layer, and this computational process is repeated p times to obtain the final output h^{p}_{i}, where each layer consists of multiple heads of attention, residual connectivity and normalization.

For the multi-head attention mechanism layer, the query vectorq_i, key vectork_i, and value vector v_i of each node are first computed from the learnable weight matrices $W^{Q'}, W^{K'}, W^{V'}$:

$$q_i = W^{Q'} h^{p-1}_{i}, k_i = W^{K'} h^{p-1}_{i}, v_i = W^{V'} h^{p-1}_{i}, i \in \{1, \ldots, n\} \tag{17}$$

Then, the attention coefficient between nodes j and $j\prime$ is calculated:

$$u_{ii'} = softmax\left(\frac{q^{T}_{i} k_{i'}}{\sqrt{d_{q'}}}\right) \tag{18}$$

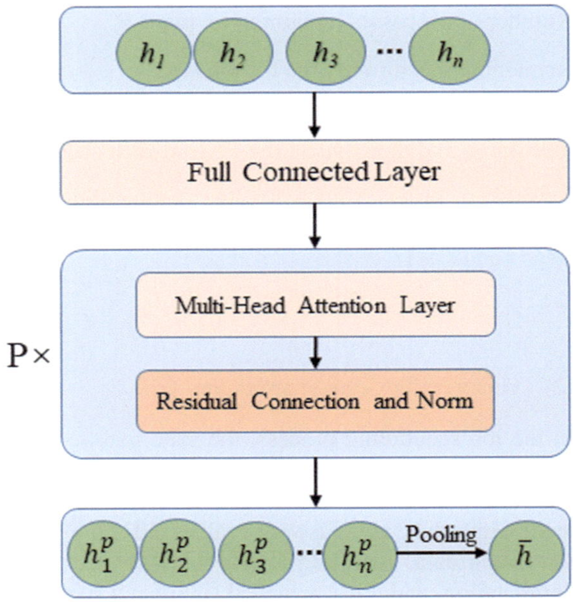

Fig. 3 The encoder structure of TSA

where q_j^T is the transpose of q_j and $d_{q'}$ is the first dimension of $\mathrm{W}^{Q'}$.

We compute the attention coefficients HHH times, then multiply them with the value vectors and concatenate the results. After that, the concatenated output is passed through a fully connected layer to obtain the output of the multi-head attention mechanism:

$$\delta_i = A_3\left(\sum_{h=1}^{H}\sum_{i'=1}^{n}(u_{ii'})^h(v_i)^h\right) \tag{19}$$

Finally, the embedding of each job node is obtained by residual concatenation and normalization, while the embedding of all job nodes is averaged to obtain the graph embedding vector $\overline{h_i^p}$, h_i^p and $\overline{h_i^p}$ as the final output of the encoder.

$$h_i^p = BN\left(h_i^{p-1} + \delta_i\right) \tag{20}$$

$$\overline{h_i^p} = \frac{1}{n}\sum_{i=1}^{n} h_i^p \tag{21}$$

For the decoder, we use the same structure as the MSA's decoder above, where we directly compute the query vector by using $\overline{h_i^p}$ as the context vector, compute the key vector and the value vector by using the embedding vectors of the nodes of each job, h_i^p, and compute the probability of selecting each job by the same multi-attention mechanism, as well as the probability of selecting the completed jobs by utilizing the masking mechanism set to 0.

5 Computational Experiment

5.1 Instance Generation and Parameter Setting

A large amount of training data is required in the training phase of the model, so we train by randomly generated datasets, for small scale arithmetic generate four sets of arithmetic with n × m × u of 8 × 4 × 2, 8 × 8 × 2, 10 × 5 × 2, and 20 × 10 × 2, in which 1600 instances are generated for each scale for training, and 128 instances are used for performance testing. For larger scale problems, arithmetic cases of sizes 30 × 20 × 5, 50 × 20 × 10, and 100 × 20 × 10 were generated, and due to the sharp increase in training time due to the increase in the number of arithmetic sequences, we generated only 800 instances for the 30 × 20 × 5 size problem for training and tested the trained model on 128 instances of the three scales, and also verified the generalization performance of the model.

In the critical parameter settings, for different problem sizes, we used the same parameters. In the training of MSA, the batch size is 16, the embedding dimension is 128, the number of heads in the decoder's multi-head attention mechanism is 8, and the Adam optimizer is used with a learning rate of $lr = 1 \times 10^{-3}$. In the TSA training, the batch size is 16, the embedding dimension is 128, the number of heads in the transformer structure's multi-head attention mechanism is 8, and the PPO algorithm's clipping coefficient ϵ, value function coefficient c_1, and policy entropy coefficient c_2 are set to 0.2, 1, and 0.01, respectively, using the Adam optimizer with a learning rate of $lr = 1 \times 10^{-3}$.

5.2 Ablation Experiments for MSA and TSA

In order to verify the effectiveness of the proposed MSA and TSA in FJSPT, we tested them on randomized instances of different sizes after performing individual training and compared them with several scheduling rules. For machine selection on the test benchmarks, we used three scheduling rules: (1) shortest processing time (SPT); (2) shortest transportation time (STT); and (3) shortest service time (SST). For job selection the same three scheduling rules are used as comparisons: (1) first-in-first-out (FIFO); (2) longest waiting time (LWT); and (3) maximum remaining time (MWKR). In particular, when conducting separate tests for MSA and TSA, the remaining scheduling tasks are completed using FIFO and SPT, respectively, and the test results are shown in Tables 2 and 3. We use the maximum completion time as the evaluation metric and compute the gap value, and we can see that our proposed MSA has a significant advantage due to all the scheduling rules, especially on larger-scale instances. In addition, on the test of $50 \times 20 \times 10$, $100 \times 20 \times 10$ scale instances, we used the model trained from $30 \times 20 \times 5$ scale instances, and it can be seen that the model has good generalization performance.

Table 2 MSA test results

Size (*n*-*m*-*v*)		SPT	STT	SST	MML	MSA (ours)
8-4-2	Cmax	128.59	230.88	135.41	154.91	118.28
	Gap	8.72%	95.20%	14.48%	30.97%	**0.00%**
8-8-2	Cmax	275.20	319.88	237.59	306.01	212.20
	Gap	29.69%	50.74%	11.97%	44.21%	**0.00%**
10-5-2	Cmax	211.790	301.630	189.380	243.190	176.410
	Gap	20.06%	70.98%	7.35%	37.85%	**0.00%**
20-10-2	Cmax	878.38	677.49	759.11	943.75	665.27
	Gap	32.03%	1.84%	14.11%	41.86%	**0.00%**
30-20-5	Cmax	998.00	1271.90	896.45	1016.00	647.45
	Gap	54.14%	96.45%	38.46%	56.92%	**0.00%**
50-20-10	Cmax	780.95	2083.70	699.65	867.81	499.05
	Gap	56.49%	317.53%	40.20%	73.89%	**0.00%**
100-20-10	Cmax	1543.10	3870.90	1356.80	1673.20	968.45
	Gap	59.34%	299.70%	40.10%	72.77%	**0.00%**

Table 3 TSA test results

Size (*n*-*m*-*v*)		FIFO	LWT	MWKR	MSA (ours)
8-4-2	Cmax	133.20	137.58	138.45	125.78
	Gap	5.90%	9.38%	10.07%	**0.00%**
8-8-2	Cmax	225.66	241.74	231.82	192.28
	Gap	17.36%	25.72%	20.56%	**0.00%**
10-5-2	Cmax	183.02	194.26	189.80	161.02
	Gap	13.66%	20.64%	17.87%	**0.00%**
20-10-2	Cmax	715.69	754.91	710.08	528.32
	Gap	35.47%	42.89%	34.40%	**0.00%**
30-20-5	Cmax	815.27	906.59	818.56	638.13
	Gap	27.76%	42.07%	28.27%	**0.00%**
50-20-10	Cmax	624.98	719.34	640.63	500.91
	Gap	24.77%	43.61%	27.89%	**0.00%**
100-20-10	Cmax	1228.79	1379.24	1231.16	994.12
	Gap	23.61%	38.74%	23.84%	**0.00%**

5.3 Test Results of the Centrally Trained Model

After the performance tests on the MSA and TSA, we then centrally trained the two agents, firstly using the trained model for the test of randomly generated instances and comparing it with the results obtained by a single agent.

Secondly, our model was tested on a standard dataset, test set 1 from Deroussi and Norre, consisting of ten small-scale instances, each with 5–8 jobs, 8 machines, and 2 AGVs, where each process can be machined on two machines with the same machining time. For each instance, Table 4 shows the optimal solution obtained by solving the MILP model, the results obtained by the state-of-the-art heuristic algorithm proposed by Homayouni et al. and the results obtained by our model, with the corresponding gap values calculated.

Table 4 Dataset 1 test results

Instance	Size (*n*-*m*-*v*)		MILP	BRKGA	LAHC	Ours
FJSPT 1	7-8-2	Cmax	134	138	138	144
		Gap	0	2.99%	2.99%	**7.46%**
FJSPT 2	6-8-2	Cmax	114	118	114	118
		Gap	0	3.51%	0.00%	**3.51%**
FJSPT 3	6-8-2	Cmax	120	120	120	124
		Gap	0	0.00%	0.00%	**3.33%**
FJSPT 4	5-8-2	Cmax	114	120	114	118
		Gap	0	5.26%	0.00%	**3.51%**
FJSPT 5	5-8-2	Cmax	94	96	94	96
		Gap	0	2.13%	0.00%	**2.13%**
FJSPT 6	6-8-2	Cmax	138	138	138	148
		Gap	0	0.00%	0.00%	**7.25%**
FJSPT 7	8-8-2	Cmax	110	112	112	121
		Gap	0	1.82%	1.82%	**10.25%**
FJSPT 8	6-8-2	Cmax	178	178	178	194
		Gap	0	0.00%	0.00%	**8.99%**
FJSPT 9	5-8-2	Cmax	144	144	144	152
		Gap	0	0.00%	0.00%	**5.56%**
FJSPT 10	6-8-2	Cmax	174	174	174	190
		Gap	0	0.00%	0.00%	**9.20%**

6 Conclusion

In this paper, we propose a multi-agent reinforcement learning framework to learn the scheduling strategy for the Flexible Job Shop Scheduling Problem with Transportation (FJSPT) aimed at optimizing the maximum completion time. We decompose FJSPT into three sub-problems: first, a graph neural network-based model framework and training algorithm are developed for the machine selection sub-problem; second, a transformer-based model framework and PPO-based training algorithm are designed for the job selection sub-problem. The experimental results demonstrate that the two agents, MSA and TSA, perform well, significantly outperforming all conventional scheduling rules. After centralized training, the framework's performance is comparable to that of advanced heuristic scheduling rules, and it outperforms various heuristic algorithms in terms of solving time for large-scale instances. It effectively balances solution quality and computational efficiency.

Although our proposed framework achieves good results in testing, it still has some limitations. For instance, the training time and memory usage increase significantly as the problem size grows, making it challenging to train on larger-scale problems. Optimizing the training process for the two agents remains a key challenge for future work. Additionally, our current model does not account for dynamic perturbations, such as machine failures or order insertion and withdrawal, which can impact scheduling decisions. Moreover, we focus solely on minimizing completion time in our optimization objective, without considering other indicators like delivery time or energy consumption. Exploring dynamic scheduling in FJSPT and optimizing for multiple objectives are important directions for our future research.

References

Bengio Y, Lodi A, Prouvost A (2021) Machine learning for combinatorial optimization: a methodological tour d'horizon. Eur J Oper Res 290(2):405–421

Berterottière L, Dauzère-Pérès S, Yugma C (2023) Flexible job-shop scheduling with transportation resources. Eur J Oper Res S0377221723005969

Chaudhry IA et al (2022) Integrated scheduling of machines and automated guided vehicles (AGVs) in flexible job shop environment using genetic algorithms. Int J Ind Eng Comput 13(3):343–362. https://doi.org/10.5267/j.ijiec

Chen R, Li W, Yang H (2023) A deep reinforcement learning framework based on an attention mechanism and disjunctive graph embedding for the job-shop scheduling problem. IEEE Trans Industr Inf 19(2):1322–1331

Dai M, Tang D, Giret A, Salido MA (2019) Multi-objective optimization for energy-efficient flexible job shop scheduling problem with transportation constraints. Robot Comput-Integr Manuf 59:143–157

Dai H, Khalil EB, Zhang Y, Dilkina B, Song L (2018) Learning combinatorial optimization algorithms over graphs. [cs, stat]. http://arxiv.org/abs/1704.01665. Last Accessed 13 Dec 2021

Deroussi L, Norre S (2010) Simultaneous scheduling of machines and vehicles for the flexible job shop problem

Du Y, Li J, Li C, Duan P (2022) A reinforcement learning approach for flexible job shop scheduling problem with crane transportation and setup times. IEEE Trans Neural Netw Learn Syst 1–15
Gao K, Cao Z, Zhang L, Chen Z, Han Y, Pan Q (2019) A review on swarm intelligence and evolutionary algorithms for solving flexible job shop scheduling problems. IEEE/CAA J Autom Sinica 6(4):904–916
Homayouni SM, Fontes DBMM (2021) Production and transport scheduling in flexible job shop manufacturing systems. J Glob Optim 79(2):463–502
Homayouni SM, Fontes DBMM, Gonçalves JF (2023) A multistart biased random key genetic algorithm for the flexible job shop scheduling problem with transportation. Int Trans Operational Res 30(2):688–716
Hu L, Liu Z, Hu W, Wang Y, Tan J, Wu F (2020) Petri-net-based dynamic scheduling of flexible manufacturing system via deep reinforcement learning with graph convolutional network. J Manuf Syst 55:1–14
Huang X, Yang L (2019) A hybrid genetic algorithm for multi-objective flexible job shop scheduling problem considering transportation time. IJICC 12(2):154–174
Kumar MVS, Janardhana R, Rao CSP (2011) Simultaneous scheduling of machines and vehicles in an FMS environment with alternative routing. Int J Adv Manuf Technol 53(1–4):339–351
Lacomme P, Larabi M, Tchernev N (2013) Job-shop based framework for simultaneous scheduling of machines and automated guided vehicles. Int J Prod Econ 143(1):24–34
Lee YH, Lee S (2022) Deep reinforcement learning based scheduling within production plan in semiconductor fabrication. Expert Syst Appl 191:116222
Lei K et al (2022a) A multi-action deep reinforcement learning framework for flexible Job-shop scheduling problem. Expert Syst Appl 205:117796
Lei K, Guo P, Wang Y, Wu X, Zhao W (2022b) Solve routing problems with a residual edge-graph attention neural network. Neurocomputing 508:79–98
Lei K, Guo P, Wang Y, Zhang J, Meng X, Qian L (2024) Large-scale dynamic scheduling for flexible job-shop with random arrivals of new jobs by hierarchical reinforcement learning. IEEE Trans Industr Inf 20(1):1007–1018
Li Y, Gu W, Yuan M, Tang Y (2022) Real-time data-driven dynamic scheduling for flexible job shop with insufficient transportation resources using hybrid deep Q network. Robot Comput-Integr Manuf 74:102283
Liu R, Piplani R, Toro C (2022) Deep reinforcement learning for dynamic scheduling of a flexible job shop. Int J Prod Res 60(13):4049–4069
Luo S (2020) Dynamic scheduling for flexible job shop with new job insertions by deep reinforcement learning. Appl Soft Comput 91:106208
Nguyen ND, Nguyen TT, Vamplew P, Dazeley R, Nahavandi S (2021) A prioritized objective actor-critic method for deep reinforcement learning. Neural Comput & Applic 33(16):10335–10349
Nouri HE, Belkahla Driss O, Ghédira K (2016) Simultaneous scheduling of machines and transport robots in flexible job shop environment using hybrid metaheuristics based on clustered holonic multiagent model. Comput & Ind Eng 102:488–501
Pan Z, Wang L, Wang J, Lu J (2023) Deep reinforcement learning based optimization algorithm for permutation flow-shop scheduling. IEEE Trans Emerg Top Comput Intell 7(4):983–994
Ren W, Yan Y, Hu Y, Guan Y (2022) Joint optimisation for dynamic flexible job-shop scheduling problem with transportation time and resource constraints. Int J Prod Res 60(18):5675–5696
Song W, Chen X, Li Q, Cao Z (2023) Flexible job-shop scheduling via graph neural network and deep reinforcement learning. IEEE Trans Industr Inf 19(2):1600–1610
Vaswani A, Shazeer N, Parmar N, Uszkoreit J, Jones L, Gomez AN, Kaiser Ł, Polosukhin I (2017) Attention is all you need. In: Advances in neural information processing systems
Veličković P, Cucurull G, Casanova A, Romero A, Liò P, Bengio Y (2018) Graph attention networks. http://arxiv.org/abs/1710.10903. Last Accessed 21 Feb 2024
Wang Y-F (2020) Adaptive job shop scheduling strategy based on weighted Q-learning algorithm. J Intell Manuf 31(2):417–432

Wang H, Cheng J, Liu C, Zhang Y, Hu S, Chen L (2022) Multi-objective reinforcement learning framework for dynamic flexible job shop scheduling problem with uncertain events. Appl Soft Comput 131:109717

Yan J, Liu Z, Zhang C, Zhang T, Zhang Y, Yang C (2021) Research on flexible job shop scheduling under finite transportation conditions for digital twin workshop. Robot Comput-Integr Manuf 72:102198

Zhang Z-Q, Wu F-C, Qian B, Hu R, Wang L, Jin H-P (2023) A Q-learning-based hyper-heuristic evolutionary algorithm for the distributed flexible job-shop scheduling problem with crane transportation. Expert Syst Appl 234:121050

Industry 4.0 Hyper-Factory Planning, Scheduling, and Execution with Real-Time Data

Mingxing Li, Qu Zhou, Ting Qu, Ming Li, Haoran Ding, Shiquan Ling, and George Q. Huang

Abstract Scholars and practitioners continue to be challenged by manufacturing complexity and uncertainty, particularly in nowadays rapidly evolving markets characterized by diverse customer demands and expanding product lines. The advent of Industry 4.0 technologies significantly mitigates these challenges by enhancing transparency and traceability through real-time data integration across physical and digital realms. However, the decision-making frameworks inherited from the previous industrial revolution for Planning, Scheduling, and Execution (PSE) are ill-equipped to leverage Industry 4.0 real-time data for overcoming the challenges of complexity and uncertainty. This study proposes a novel Industry 4.0 PSE methodology, Hyper-Factory Planning, Scheduling, and Execution (HF-PSE), designed to enable adaptive and robust decision-making amidst diverse uncertainties in contemporary manufacturing settings. HF-PSE integrates the fundamental spirit and adaptable principles of university graduation ceremonies into the production management process, including clustering-based planning with a rolling horizon, real-time scheduling based on workstation-job ticket suitability, and Out-of-Order ticketing for production execution. A case study demonstrates that HF-PSE surpasses others on average and exhibits

M. Li (✉) · Q. Zhou · T. Qu
School of Intelligent Systems Science and Engineering, Jinan University, Zhuhai, China
e-mail: limingxing@jnu.edu.cn

Guangdong International Cooperation Base of Science and Technology for GBA Smart Logistics, Jinan University, Zhuhai, China

Institute of Physical Internet, Jinan University, Zhuhai, China

M. Li · G. Q. Huang
Department of Industrial and Systems Engineering, The Hong Kong Polytechnic University, Hung Hom, Hong Kong, China

H. Ding
Department of Data and Systems Engineering, The University of Hong Kong, Hong Kong, China

S. Ling
School of Mechanical and Electrical Engineering, Shenzhen Polytechnic University, Shenzhen, China

S. Huang et al. (eds.), *Design and Operation of Smart Reconfigurable Manufacturing Systems in Industry 4.0/5.0*, Springer Series in Advanced Manufacturing,
https://doi.org/10.1007/978-3-032-00284-6_8

minimal statistical variations in cost-efficiency, punctuality, and simultaneity metrics, highlighting its superior effectiveness, stability, and resilience in stochastic settings.

1 Introduction

Production planning, scheduling, and execution (PSE) encompass the routine decision-making and implementation processes for allocating and utilizing limited resources to meet demand in the manufacturing sector (Pinedo 2005; Allahverdi et al. 2008; Li et al. 2022a). PSE are pivotal in contemporary production management, attracting significant interest from both academia and industry over the past decades. Researchers are particularly intrigued by PSE challenges and models, especially the planning and scheduling aspects, due to their theoretical complexity and practical significance (Pinedo 2016). Practitioners focus on creating efficient tools to optimize PSE processes in factories, aiming for simultaneous enhancements in productivity, cost-effectiveness, and customer satisfaction (Tenhiälä 2011).

Numerous sophisticated mathematical models and intricate algorithms have been proposed over the past decades to address practical PSE challenges (Allahverdi et al. 2008; Jiang et al. 2022; Li et al. 2022b; Mula et al. 2006). Mathematical programming techniques like integer programming and dynamic programming are favoured for formulating PSE problems (Allahverdi et al. 2008; Li et al. 2022b). Subsequently, exact algorithms can optimally solve small-scale problems (Brah and Hunsucker 1991; Magnanti and Vachani 1990). Heuristics (Đurasević and Jakobović 2018) and meta-heuristics (Li et al. 2022a) approaches are proposed to efficiently address large-scale PSE problems, achieving near-optimal performance with reasonable computational costs. Theoretical research in production PSE has advanced swiftly, emphasizing optimality and innovative methodologies. Nevertheless, reports indicate that many planners and schedulers still rely on simple, flexible tools like spreadsheets for routine tasks (de Man and Strandhagen 2018), and shy away from the complex, theoretically optimal or near-optimal methods explored in the literature (Tenhiälä 2011; Oluyisola et al. 2021). A significant disparity exists between PSE theory and practice, primarily because real-world PSE problems are challenging to model accurately (Chakhlevitch and Cowling 2008), numerous assumptions are required, often neglecting real-world randomness. These mathematically “optimal” solutions may not be the best choices in practice (Chakhlevitch and Cowling 2008). Adaptable and robust PSE principles are essential to bridging the gap between theory and practice.

Moreover, uncertainties, disturbances, and disruptions frequently occur in manufacturing systems, inevitably undermining system performance and stability (Li et al. 2023a, 2023b; Elyasi et al. 2024; Liu et al. 2017; Ghaleb et al. 2020). These unpredictable events significantly contribute to the discrepancies between production PSE theory and practice. Thus, integrating uncertainties into PSE models and methods is crucial for Industry 4.0 manufacturing (Guo et al. 2021). Stochastic processing time represents a common form of uncertainty in manufacturing systems (Laslo et al. 2008; Hazır and Dolgui 2013). Moreover, researchers frequently account for machine unavailability, failure, or breakdown as another prevalent uncertainty (Bi et al. 2024; Mirabi et al. 2013; Zandieh et al. 2017). Subsequently, joint production-maintenance scheduling issues have begun to garner attention (Lu et al. 2015). In recent years, more intricate combinations of various uncertainties have been investigated (Liu et al. 2017; Ghaleb et al. 2020). Nevertheless, the majority of existing studies on manufacturing uncertainties have been specifically designed for one or two particular types. Typically, probability distributions are presumed to be known or predictable. However, real-world manufacturing systems are influenced by a combination of both predictable and unpredictable uncertainties. Industry 4.0 manufacturing urgently requires a comprehensive approach that can manage diverse uncertainties with flexibility and resilience.

Furthermore, for decades, traditional PSE modes and models have dominated manufacturing systems, even as technology has advanced rapidly. Emerging concepts and technologies, such as Cyber-Physical Systems (CPS), digital twins (DT), and the industrial Internet of Things (IIoT), are profoundly transforming contemporary manufacturing landscapes. Technological advancements from the industrial revolution have made it feasible to collect vast amounts of manufacturing data in real-time from diverse sources, with enhanced reliability and quality (Li and Huang 2021; Tao et al. 2018). Scholars have begun employing data-driven approaches for production PSE (Hubbs et al. 2020; Demirhan et al. 2020; Li et al. 2023c, 2024). Nevertheless, the majority of earlier data-driven approaches are heavily dependent on high-quality training datasets and big data. Few studies have explored utilizing real-time data to support immediate decision-making in the PSE process, aiming to enhance overall operational resilience amid uncertainties.

This study seeks to develop a comprehensive Industry 4.0 PSE framework, Hyper-Factory Planning, Scheduling, and Execution (HF-PSE), integrating the fundamental principles and adaptable concepts of university graduation ceremonies into production management. This includes clustering-based planning with a rolling horizon, real-time scheduling based on workstation-job ticket suitability, and out-of-order ticketing for production execution. HF-PSE decisions are highly adaptable, allowing for real-time modifications to address stochastic elements. Real-time data is incorporated into the HF-PSE model to ensure decisions align with the current shop floor status, thereby enhancing overall resilience.

The remainder of this article is organized as follows. Section 2 provides an overview of HF-PSE and graduation ceremonies. Section 3 introduces the general procedure and mathematical models of HF-PSE. Section 4 explores the application of HF-PSE in a fixed-position assembly workshop and conducts a numerical analysis to evaluate its performance. Section 5 concludes with a summary of findings and future directions.

2 The Overview of Hyper-Factory Planning, Scheduling, and Execution

The Hyper-Factory Planning, Scheduling, and Execution (HF-PSE) system is inspired by the core spirit and principles of university graduation ceremonies. The effectiveness of these ceremonies is primarily due to ticket-based adaptive management and the seamless real-time coordination among participants. HF-PSE integrates these adaptive and resilient strategies, employing Industry 4.0 technologies to enhance manufacturing efficiency.

Figure 1 demonstrates the alignment between the HF-PSE framework and a graduation ceremony, emphasizing their direct correspondence. Four types of functional tickets streamline the workflow: Intra-Logistics Tickets (LTs)/student lists, Job Tickets (JTs)/admission tickets, Setup Tickets (STs)/program tickets, and Operation Tickets (OTs)/name tickets. The factory dynamically receives orders/jobs (similar to annual student enrolment); Jobs/orders are categorized and dispatched to the shop floor, with LTs generated automatically based on material needs (as students are grouped by programs/majors, defining credit requirements); The intralogistics department prepares materials according to LTs (like students attending courses with credits and grades recorded in lists); Jobs are confirmed for production upon resource validation, triggering JTs issuance (as students meeting graduation criteria receive admission tickets); Validated jobs queue on the shop floor for sequential processing (like students lining up at the ceremony stage); When an ST enters the queue, workstation setup is pre-emptively informed (similar to notifying department heads and principals of program changes); Operations proceed sequentially per OTs (as recipients ascend the stage, present name cards, have their names announced, receive degrees, and descend).

The core principle of graduation ceremonies is managing disruptions by adhering to foundational guidelines, a notion equally relevant to manufacturing systems management. However, several challenges arise. First, how can we evaluate similarity and classify analogous orders or tasks during production planning? Second, how should JTs be prioritized within the JT queue, and how can their sequence be dynamically adjusted based on real-time conditions during production scheduling? Lastly, considering the complexity of manufacturing operations, how can the OT queue be organized to ensure seamless execution and enhanced efficiency?

3 Mathematical Formulation of Hyper-Factory Planning, Scheduling, and Execution

This section mathematically formulates the HF-PSE procedure, integrating clustering-based rolling horizon planning to categorize similar orders or jobs, real-time JT scheduling based on workstation-job ticket suitability, and Out-of-Order

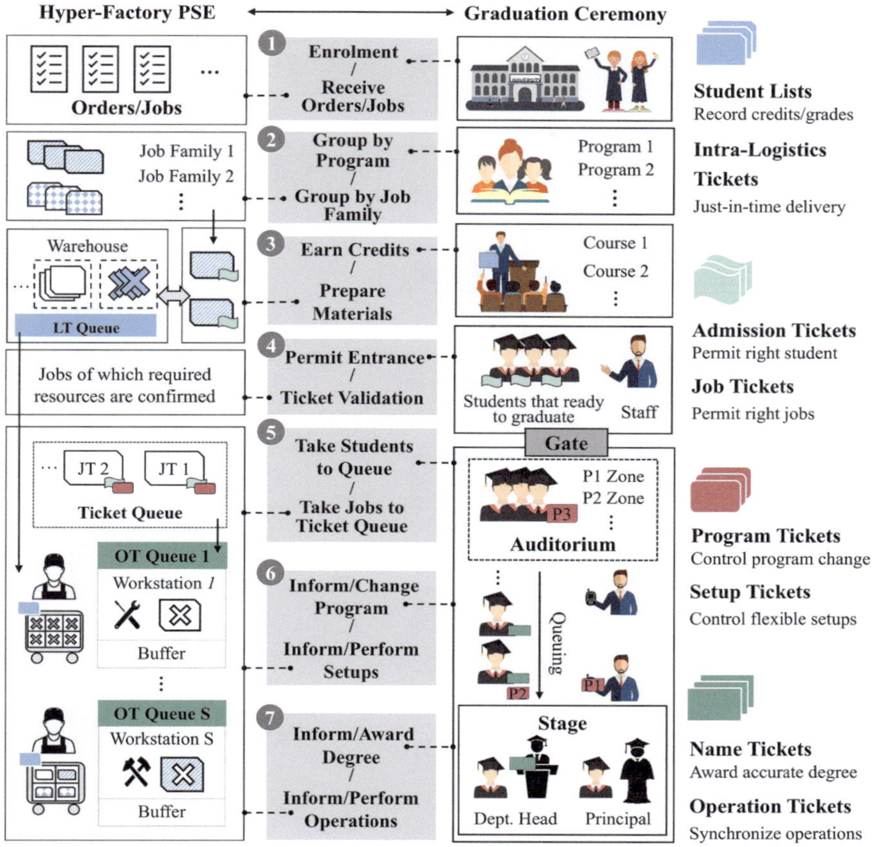

Fig. 1 The overview of Hyper-Factory PSE and graduation ceremony

ticketing for production execution, as depicted in Fig. 2. The HF-PSE flowchart is presented in Fig. 3, and the notations used are listed in Table 1.

3.1 Clustering-Based Planning with Rolling Horizon

The clustering-based planning approach aggregates similar orders or jobs into clusters for production within each planning period, leveraging a rolling horizon strategy to address uncertainty. Generally, the planning process involves five steps.

Firstly, establishing the criteria for assessing the similarity of orders/JTs. The choice of suitable similarity attributes directly influences the method's performance on relevant metrics. For example, incorporating attributes related to the bill of materials can boost intralogistics efficiency. Selecting attributes linked to due dates ensures

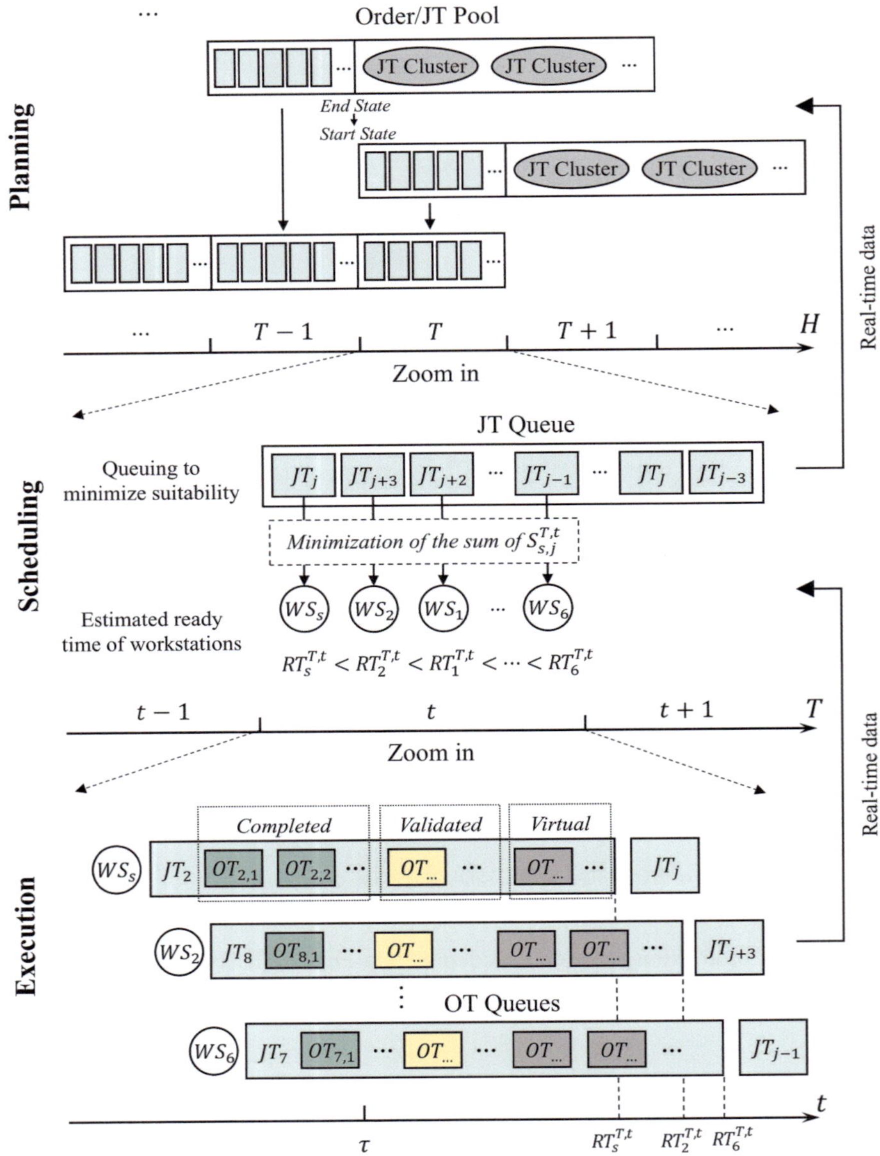

Fig. 2 Hyper-Factory Planning, Scheduling, and Execution

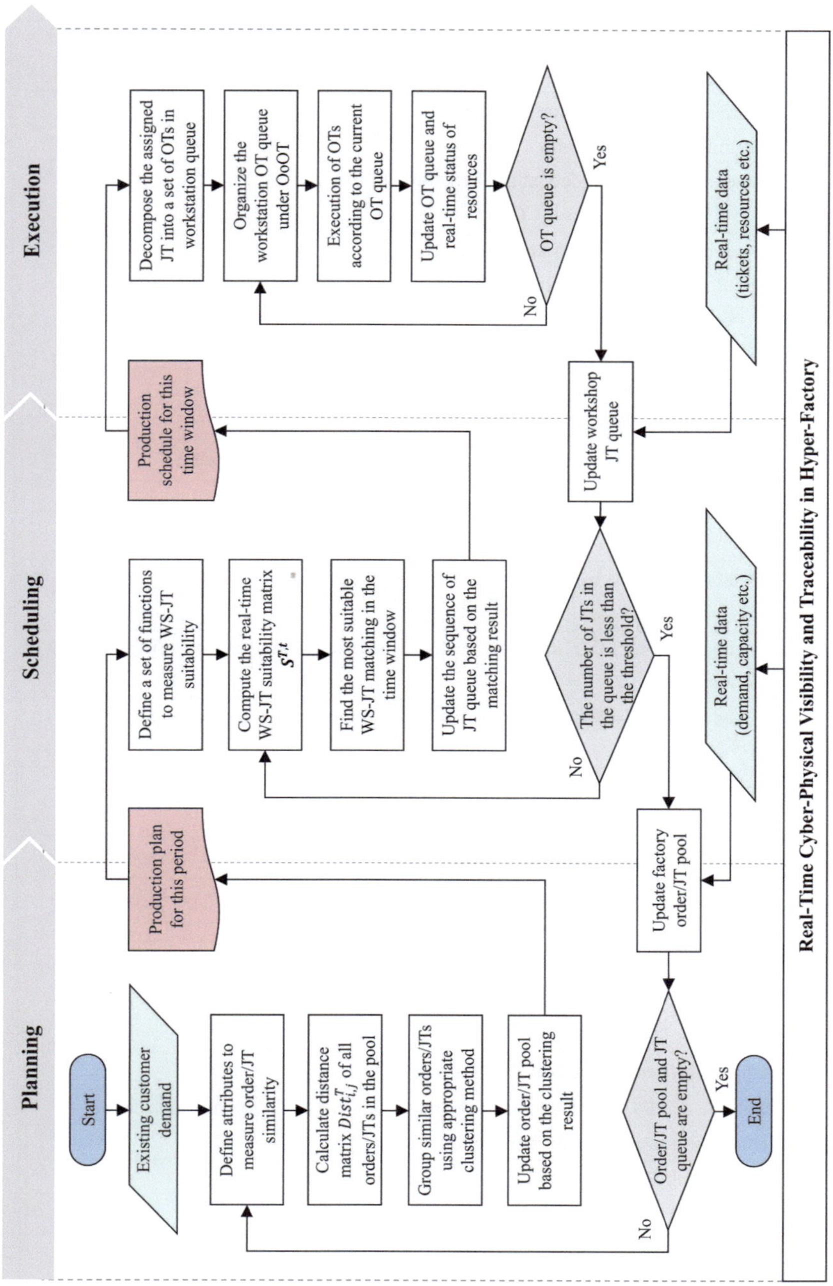

Fig. 3 The flowchart of Hyper-Factory PSE

Table 1 Notations and descriptions

Notations	Descriptions
H	The overall decision horizon, $H = \{1, 2, \ldots, T-1, T, T+1, \ldots\}$
T	Production planning time period, $T = \{1, 2, \ldots, t-1, t, t+1, \ldots\}$
t	Production scheduling time window
τ	Real-time, the current time
CO^T	The set of customer orders at the beginning of time period T, $CO^T = \{1, 2, \ldots, o, \ldots, \lvert CO^T \rvert\}$
F_o	The set of product types/families requested in order o
JP^T	JT pool, the set of JTs at the beginning of time period T, $JP^T = \{1, 2, \ldots, i, \ldots, j, \ldots, \lvert JP^T \rvert\}$
$h_{j,o}$	1 if JT j belongs to customer order o, and 0 otherwise
$J^T_{release}$	The set of JTs released at the beginning of time period T
J^T_{new}	The set of new JTs arrived during time period T
$dist^{T,a}_{i,j}$	The distance between JT i and j regarding attribute a at time period T
$Dist^T_{i,j}$	The overall distance between JT i and j at time period T, distance matrix $\boldsymbol{D^T} = (Dist^T_{i,j})_{i,j \in JP^T}$
$WS^{T,t}$	The set of available workstations at the beginning of time window t, $WS^{T,t} = \{1, 2, \ldots, s, \ldots, \lvert WS^{T,t} \rvert\}$
$JQ^{T,t}$	Job queue, the set of JT s that has been released to the shop floor at the beginning of time window t, $JQ^{T,t} = \{1, 2, \ldots, i, \ldots, j, \ldots, \lvert JQ^{T,t} \rvert\}$
$WIP^{T,t}$	The set of WIPs (work-in-progress) at the beginning of time window t, $WIP^{T,t} = \{1, 2, \ldots, i, \ldots, j, \ldots, \lvert WIP^{T,t} \rvert\}$
$FP^{T,t}$	The set of finished products/JTs at the beginning of time window t, $FP^{T,t} = \{1, 2, \ldots, i, \ldots, j, \ldots, \lvert FP^{T,t} \rvert\}$
$LF^{T,t}_{s,j}$	The suitability of JT j for workstation s from the '*Look Forward*' perspective
$LA^{T,t}_{s,j}$	The suitability of JT j for workstation s from the '*Look Around*' perspective
$LB^{T,t}_{s,j}$	The suitability of JT j for workstation s from the '*Look Backward*' perspective
$S^{T,t}_{s,j}$	The overall suitability of JT j for workstation s, suitability matrix $\boldsymbol{S^{T,t}} = (S^{T,t}_{s,j})_{s \in WS^{T,t}, j \in JQ^{T,t}}$
pt_j	Standard processing time of JT j
dt_j	Due time of JT j, which equals to the due time dt_o if JT j belongs to order o
$st_{i,j}$	Setup time for changeover between JT i and j
$j(s,t)$	The ongoing JT at workstation s in time window t
$n^{T,t}_j$	Number of JTs that belong to the same family as JT j in $JQ^{T,t}$
$rt^{T,t}_s$	Estimated ready time of workstation s in time window t (namely, the estimated completion time of the ongoing JT at the workstation)

(continued)

Table 1 (continued)

Notations	Descriptions
$C^{j,\tau}$	Constraint matrix of JT j at the current time
RTS^{τ}	**A general set that includes all the concerning real-time status**

improved timeliness in orders/JTs completion. The selection process is highly flexible, enabling decision-makers to make nuanced attribute choices. Furthermore, the unique characteristics of selected attributes must be considered when quantifying their similarity.

Secondly, computing the distance matrix for all confirmed orders/JTs in the pool. Pairwise distance is often used to quantify dissimilarity; notably, a smaller distance indicates greater similarity between two orders/JTs. Various distance calculation methods should be adapted to different attribute types. For example, binary values like 0 and 1 can efficiently represent job family similarity, where 0 indicates the same family and 1 denotes otherwise; Euclidean distance is suitable for evaluating due date similarity. The overall distance is obtained by integrating all attribute distances with appropriate weighting factors.

$$Dist_{i,j}^{T} = \sum_{a=1}^{A} w_a \cdot dist_{i,j}^{a,T} \tag{1}$$

For the JT pool JP^T consisting of $\left|JP^T\right|$ JTs, there are $\frac{\left|JP^T\right| \cdot \left(\left|JP^T\right| - 1\right)}{2}$ order/JT pairs. This results in a $\left|JP^T\right| \times \left|JP^T\right|$ distance matrix.

$$\boldsymbol{D^T} = \begin{bmatrix} 0 & Dist_{1,2}^{T} & \cdots & Dist_{1,\left|JP^T\right|}^{T} \\ Dist_{2,1}^{T} & 0 & & Dist_{2,\left|JP^T\right|}^{T} \\ \vdots & & \ddots & \vdots \\ Dist_{\left|JP^T\right|,1}^{T} & Dist_{\left|JP^T\right|,2}^{T} & \cdots & 0 \end{bmatrix} \tag{2}$$

Thirdly, using an appropriate method to cluster similar orders/JTs. The distance matrix quantifies the proximity between orders/JTs. The clustering method should effectively leverage this precomputed distance matrix. Examples include hierarchical clustering and spectral clustering, which are grounded in graph theory.

Fourthly, releasing orders/JT clusters sequentially, taking into account real-time conditions. Decision-makers have the flexibility to introduce clusters into the shop floor JT queue. The JT queue is the set of all released JTs to be processed in this period.

$$JQ^{T,1} = JQ^{T-1,end} \cup J_{release}^{T-1} \tag{3}$$

Released JTs in the queue remain stable during decision period T, avoiding return to the JT pool for further clustering, thus ensuring uninterrupted scheduling and execution. Only significant disturbances might necessitate re-evaluation, which is beyond the scope of this paper.

Finally, at the end of each period, the orders/JT pool and shop floor status are updated to reflect newly arrived orders or jobs, and any demand fluctuations during the period, with the shop floor status adjusted accordingly.

$$JP^T = \left(JP^{T-1} \backslash J_{release}^{T-1}\right) \cup J_{new}^{T-1} \tag{4}$$

3.2 Real-Time Scheduling Based on Workstation-Job Ticket Suitability

The hyper-factory production scheduling simplifies to sequencing JTs in the shop floor JT queue. Clustering ensures that released JTs over a period are somewhat similar, making rigid sequencing less critical. Moreover, rigid schedules often weaken resilience to uncertainties. Thus, this section proposes workstation-job ticket (WS-JT) suitability for flexible, resilient real-time production scheduling.

In practice, scheduling and processing a job on a specific workstation is not a binary decision; it is often more precise to consider the preference for potential assignments. The proposed scheduling method incorporates the concept of assignment preference. For each workstation s in $WS^{T,t}$, the real-time suitability of JT j is denoted by $S_{s,j}^{T,t}$, serving as a direct measure of the preference to process JT j on workstation s.

At the start of each period, prioritized clusters are released into the shop floor JT queue, where the real-time status of the queue $JQ^{T,t}$ will be monitored and updated. Available operators/workstations review the validated JT queue to identify suitable JT for processing within this time window. Suitability must be assessed from multiple perspectives.

The '*Look Forward*' index assesses the urgency of a JT and is typically expressed as a function of attributes like due date, customer priority, and related factors.

$$Look\ Forward : LF_{s,j}^{T,t} = LF_{Function\left(x_1^{LF}, x_2^{LF}, \ldots\right)}, x_1^{LF}, x_2^{LF}, \ldots \in RTS^{\tau} \tag{5}$$

The '*Look around*' index assesses the onsite conditions, considering unreleased job tickets, work-in-progress levels, finished products, and related factors.

$$Look\ Around : LA_{s,j}^{T,t} = LA\ Function\left(x_1^{LA}, x_2^{LA}, \ldots\right), x_1^{LA}, x_2^{LA}, \ldots \in RTS^{\tau} \tag{6}$$

The '*Look backward*' index typically considers the status within the previous time window, such as the setup conditions of the selected workstation.

$$Look\ Backward : LB_{s,j}^{T,t} = LB\ Function\left(x_1^{LB}, x_2^{LB}, \ldots\right), x_1^{LB}, x_2^{LB}, \ldots \in RTS^{\tau} \quad (7)$$

It is important to highlight that the suitability calculation is extremely adaptable and can be tailored to specific needs and circumstances. This section merely provides a general guideline. The overall suitability $S_{s,j}^{T,t}$ must balance various evaluation criteria, with w_{lf}, w_{la}, and w_{lb} denoting the weights.

$$S_{s,j}^{T,t} = w_{lf} \cdot norm\left(LF_{s,j}^{T,t}\right) + w_{la} \cdot norm\left(LA_{s,j}^{T,t}\right) + w_{lb} \cdot norm\left(LB_{s,j}^{T,t}\right) \quad (8)$$

The suitability matrix $\boldsymbol{S^{T,t}}$, with dimensions $\left|WS^{T,t}\right| \times \left|JQ^{T,t}\right|$, is obtained as follows.

$$\boldsymbol{S^{T,t}} = \begin{bmatrix} S_{1,1}^{T,t} & S_{1,2}^{T,t} & \cdots & S_{1,j}^{T,t} & \cdots & S_{1,\left|JQ^{T,t}\right|}^{T,t} \\ S_{2,1}^{T,t} & S_{2,2}^{T,t} & & S_{2,j}^{T,t} & & S_{2,\left|JQ^{T,t}\right|}^{T,t} \\ \vdots & & & & & \vdots \\ S_{s,1}^{T,t} & S_{s,2}^{T,t} & & \ddots & & S_{s,\left|JQ^{T,t}\right|}^{T,t} \\ \vdots & & & & & \vdots \\ S_{\left|WS^{T,t}\right|,1}^{T,t} & S_{\left|WS^{T,t}\right|,2}^{T,t} & \cdots & S_{\left|WS^{T,t}\right|,j}^{T,t} & \cdots & S_{\left|WS^{T,t}\right|,\left|JQ^{T,t}\right|}^{T,t} \end{bmatrix} \quad (9)$$

In the suitability matrix$\boldsymbol{S^{T,t}}$, row s functions as a provisional JT queue for workstation s.The suitability metrics in each row indicate the local production preferences of the associated workstation. Theoretically, workstation s selects the first JT in its queue when it becomes available for a new task. In practice, a specific JT might be prioritized by multiple workstations, and assigning the first JT to a particular workstation can cause significant rearrangements in other temporary JT queues. Thus, a coordination mechanism is essential among workstations for WS-JT alignment within this scheduling window.

The WS-JT matching within a scheduling window can be mathematically conceptualized as finding permutations of the suitability matrix $\boldsymbol{S^{T,t}}$ to minimize diagonal elements, akin to the minimum weighted bipartite matching problem in graph theory (Duff and Koster 2001). Specifically, rows can be reordered based on the estimated readiness of workstations $WS^{T,t}$, with the earliest available workstation placed in the first row. The task then is to determine column permutations (ordering the JTs in $JQ^{T,t}$) to minimize the sum of diagonal suitability indices, achieving the optimal WS-JT matching within the time window. JTs are assigned to workstations according to this matching result.

3.3 Out-of-Order Ticketing for Execution

Upon allocation to a workstation, a JT is decomposed into a series of OT s, each representing a specific operation outlined in process planning, which determines the sequence of operations necessary for producing a part or assembling a product. Some process plans are straightforward with uniquely sequenced operations, while others are more intricate, offering alternative operations. The execution of manufacturing processes in real-world settings is influenced by various uncertainties, such as stochastic processing times and machine/tool breakdowns, due to the involvement of multiple diverse resources. Therefore, a straightforward and resilient strategy is essential to manage the operation execution process, ensuring smooth workflow and improved resource utilization under uncertainty.

The concept of Out-of-Order ticketing (OoOT) draws inspiration from the Out-of-Order execution (OoOE) logic found in modern computer CPUs. OoOE arranges the sequence of pending instructions based on the availability of necessary input data and corresponding functional units, rather than their original order in the program, thereby minimizing stalls and delays. OoOT adopts this flexible processing logic to oversee the production execution process, enabling operators to make adaptive decisions according to real-time conditions.

The OT queues are typically categorized into three segments: virtual OT s, validated OT s, and completed OT s. Initially, the virtual OT queue contains all operations of the released JT. Once necessary resources are verified through shop floor visibility and traceability, and production constraints like precedence are satisfied, the OT transitions to the validated OT queue, indicating it is ready for execution.

Operators execute production tasks sequentially based on the OT queues, with all pertinent data and information updated in real-time. For example, operational precedence constraints are represented by a constraint matrix $\boldsymbol{C}^{j,\tau}$, where $C_{1,2}^{j,\tau} = 1$ indicates $OT1$ must precede $OT2$, and $C_{1,2}^{j,\tau} = 0$ signifies that $OT1$ does not need to precede $OT2$ or $OT1$ has already been completed. Upon completion of an OT, the entries in the corresponding row are set to zero. The satisfaction of an OT's precedence constraints is recorded by these row entries. Thus, the operational constraints of OT queues are straightforwardly documented and can be updated in real-time.

$$\boldsymbol{C}^{j,\tau} = \begin{bmatrix} 0 & C_{1,2}^{j,\tau} & \cdots & C_{1,P}^{j,\tau} \\ C_{2,1}^{j,\tau} & 0 & & C_{2,P}^{j,\tau} \\ \vdots & & \ddots & \vdots \\ C_{P,1}^{j,\tau} & C_{P,2}^{j,\tau} & \cdots & 0 \end{bmatrix} \tag{10}$$

Additionally, there are various statuses concerning human operators, machines, materials, and tickets, such as the estimated readiness time of these resources and their current setup conditions. A comprehensive set RTS^{τ} is established to encompass all relevant real-time statuses.

$$RTS^{\tau} = (Human, Machine, Material, Ticket, \ldots) \tag{11}$$

The real-time status RTS^{τ} captured during the execution process is fed back into the suitability evaluation within the scheduling phase and even utilized for clustering in the planning phase. This ensures that the gap between planning/scheduling and actual execution is seamlessly bridged, as decisions are aligned with the real-time conditions on the shop floor.

3.4 Key Characteristics of Hyper-Factory PSE

The mathematical formulation of HF-PSE significantly diverges from conventional PSE methods. It integrates manufacturing challenges and technological attributes of Industry 4.0 into its framework. Flexibility in HF-PSE is twofold: firstly, decisions at the planning, scheduling, and execution stages are adaptable, permitting real-time adjustments to accommodate dynamic shifts. Secondly, the methodology itself is sufficiently versatile to be easily tailored to specific needs and preferences.

Moreover, HF-PSE demonstrates greater resilience than traditional methods in stochastic settings due to its inherent flexibility. The decision-making process of HF-PSE is exceptionally adaptive to diverse manufacturing uncertainties, as real-time data are integrated into its formulation, enhancing both the stability and recoverability of the method. Consequently, HF-PSE is well-suited for modern manufacturing environments within the context of Industry 4.0. Table 2 outlines the key characteristics of the proposed HF-PSE.

4 Application of Hyper-Factory PSE in a Fixed-Position Assembly Workshop

This section details the application of HF-PSE in an Industrial Internet of Things (IIoT)-enabled fixed-position assembly workshop (FPAW). Section 4.1 outlines the layout of the FPAW, while Sect. 4.2 provides the distance metrics, suitability calculations, and key algorithms of HF-PSE for the FPAW scenario.

4.1 IIoT-Enabled Fixed-Position Assembly Workshop

The examined FPAW comprises multiple workstations designed for assembling large, bulky, and fragile products (Guo et al. 2020), as illustrated in Fig. 4. Most products are customized in small batches to meet specific customer demands. Consequently, the WIP/sub-assembly remains at a single workstation throughout the entire assembly

Table 2 Principal features of the Hyper-Factory PSE

	Traditional PSE	Hyper-Factory PSE
Period	Industry 3.0	Industry 4.0
Primary objectives	Productivity improvement Cost reduction	Profitability enhancement Customer satisfaction Resilience to uncertainties
Production mode	Mass production Multi-variety small-batch production Relatively long product lifecycle	Mass customization Personalization Relatively short product lifecycle
Manufacturing operations	Standard and repetitive operations and processes	Combination of standard and complex operations and processes
Management philosophy	Just-in-Time Material requirements planning Manufacturing Resource Planning (Benton and Shin 1998)	Synchronization Synchroperation (Guo et al. 2021)
Typical methodologies	Mathematical programming Simulation-based approaches Heuristic & Metaheuristic (Pinedo 2009)	Artificial intelligence Mathematical programming Real-time data-driven approaches (Qian et al. 2020)
Typical performance measures	Throughput & Makespan Setup costs Due date related (i.e., lateness, tardiness) Work-in-Process (WIP) (Pinedo 2009)	Simultaneity (i.e., job/order flow time, waiting time, holding time) Punctuality (i.e., earliness, lateness, tardiness) Cost-efficiency (i.e., makespan, setup time, utilization, throughput, WIP level) Robustness & Resilience (Guo et al. 2021)
Enabling technologies	Automation Information and Communications Technology Computer-integrated manufacturing	Advanced robotics and automation Industrial Internet of Things Cyber-Physical Systems & Digital twin Advanced Computing & Data analytics (Zhou et al. 2018; Olsen and Tomlin 2020)

(continued)

Table 2 (continued)

	Traditional PSE	Hyper-Factory PSE
Manufacturing data	Incomplete, inaccurate, delayed paper-based data Islands of automation, information silo (McFarlane et al. 2003; Zhong et al. 2013)	Multi-source, multi-dimension, reliable real-time data Seamless data and information sharing Data-driven decision-making (Li and Huang 2021; Tao et al. 2018)
Applicability	Usually tailored for a specific manufacturing system configuration, can seldom prevail in other environments (need reformulation)	Broad applicability for various manufacturing system configurations with minor modification
Response to uncertainties	Decisions highly rely on human experience	Decisions driven by real-time data

process. The assembly process involves a sequence of operations constrained by complex interdependencies and requiring various production resources. All resources are scheduled to move to designated islands to perform specific tasks. In an IIoT-enabled FPAW, smart IoT devices are deployed to enhance key production resources with sensing, communication, and networking capabilities.

While it is widely acknowledged that IIoT solutions improve factory connectivity and that real-time IIoT data aid production PSE, most manufacturers, given the investment in these solutions, seek more than just visibility and traceability. The core issue is the absence of methodologies to fully exploit the potential of cutting-edge technologies for supporting production PSE decision-making in the Industry 4.0 era.

4.2 *HF-PSE for Fixed-Position Assembly Workshop*

In accordance with the guidelines outlined in Sect. 3, this subsection utilizes HF-PSE within the FPAW context to aid real-time production management decisions.

In clustering-based planning, customer order similarity is assessed by product types/families and order due dates. The distance for product families $dist_{o,o'}^{f,T}$ is computed using the Jaccard distance, while the due date distance $dist_{o,o'}^{d,T}$ is determined via the Standardized Euclidean distance. The standard deviation is denoted as $std(dt)$.

$$dist_{o,o'}^{f,T} = 1 - \frac{|F_o \cap F_{o'}|}{|F_o \cup F_{o'}|} \tag{12}$$

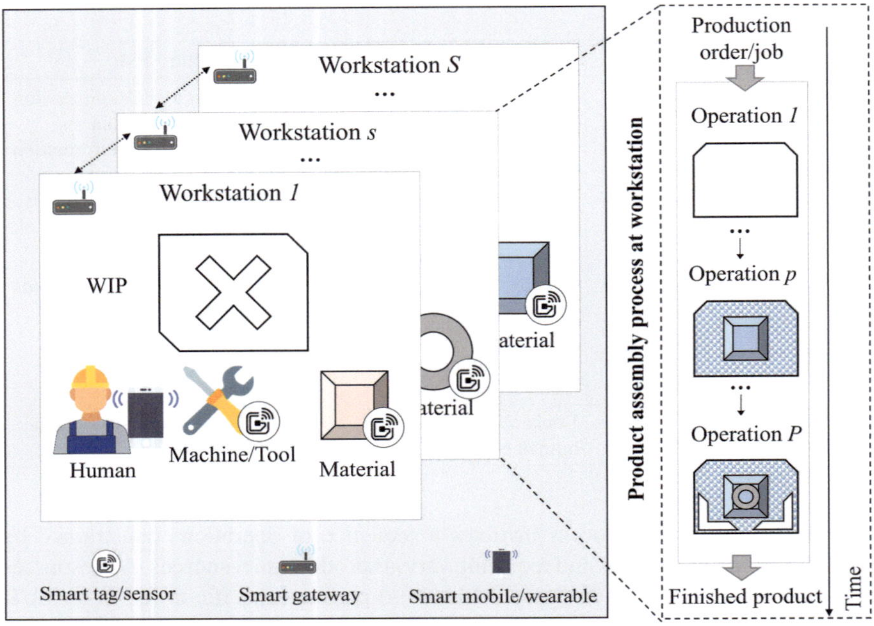

Fig. 4 IIoT-enabled fixed-position assembly workshop

$$dist_{o,o'}^{d,T} = \sqrt{\left(\frac{dt_o - dt_{o'}}{std(dt)}\right)^2} \tag{13}$$

A graph-theory-based clustering method is employed to categorize similar orders using pairwise distances, with the algorithm refined from prior research (Nascimento and De Carvalho 2011).

In real-time scheduling, the $WS \sim JT$ suitability calculation uses the '*Look Forward*' approach, indicating that a task's urgency intensifies as its deadline nears.

$$Look\ Forward : LF_{s,j}^{T,t} = dt_j - st_{j(s,t),j} - pt_j - rt_s^{T,t} \tag{14}$$

'*Look Around*' strategy guarantees seamless workflow between adjacent workstations. The metric $LA^{T,t}_{ws,j}$ evaluates the balance of WIP level and part availability at subsequent stations to minimize potential delays.

$$Look\ Around : LA^{T,t}_{s,j} = e^{\left(\frac{JQPT_O - WIPPT_O - FPPT_O}{COPT_O}\right)} \cdot r^{T,t}_{s} \tag{15}$$

where $JQPT_o$, $WIPPT_o$, $FPPT_o$, and $COPT_o$ denote the queuing work, work in progress, finished work, and total work of order o. The readiness of workstations is indicated by $r^{T,t}_{s}$. Jobs from orders with a higher completion rate are typically assigned to the earliest available workstations, and vice versa. ε is a small constant to prevent division by zero.

$$JQPT_o = \sum_{j \in JQ^{T,t}} h_{j,o} \cdot pt_j \tag{16}$$

$$WIPPT_o = \sum_{j \in WIP^{T,t}} h_{j,o} \cdot pt_j \tag{17}$$

$$FPPT_o = \sum_{j \in FP^{T,t}} h_{j,o} \cdot pt_j \tag{18}$$

$$COPT_o = \sum_{j \in JP^{T,t}} h_{j,o} \cdot pt_j \tag{19}$$

$$r^{T,t}_{s} = \begin{cases} \frac{rt^{T,t}_{s}}{\max_{s \in WS^{T,t}} \left(rt^{T,t}_{s}\right) + \varepsilon}, & ifFPPT > 0 \\ 1, & ifWIPPT > 0andFPPT = 0 \\ \frac{\max_{s \in WS^{T,t}} \left(rt^{T,t}_{s}\right)}{rt^{T,t}_{s} + \varepsilon}, & ifWIPPT = 0andFPPT = 0 \end{cases} \tag{20}$$

'*Look Backward*' seeks to enhance operator efficiency by minimizing setup tasks.

$$Look\ Backward : LB^{T,t}_{s,j} = \begin{cases} n^{T,t}_{j}, & if\ setup\ is\ not\ reuired \\ st_{j(s,t-1),j}, & if\ setup\ is\ reuired \end{cases} \tag{21}$$

The WS-JT matching algorithm is adapted from Duff and Koster (2001) to address potential mismatches.

4.3 Numerical Experiment Design

In this numerical study, the dataset is randomly generated, incorporating three typical manufacturing uncertainties: new customer order arrivals, stochastic processing times, and assembly station failures. Details are provided in Tables 3 and 4. Some probability distributions are consistent with those in prior studies (Guo et al. 2020). All numerical experiments are executed using custom MATLAB code on a personal computer (Intel Core i7-6700, x64, 3.40 GHz, 16 GB RAM).

Three other methods proposed in the literature are adopted as references in the experiments. The methods include modified earliest due date (MEDD), job with similar setup and modified earliest due date (JMEDD), and Just-in-Time (JIT), as detailed in Table 5 The MEDD and JIT are anticipated to excel in punctuality (Đurasević and Jakobović 2018; Kanet and Li 2004), while JMEDD is deemed effective for minimizing makespan, setup, and flow times (Sharma and Jain 2015). Furthermore, these methods are adaptable and dynamic for stochastic manufacturing environments. Notably, JMEDD's priority P_1 pertains to a specific workstation; thus,

Table 3 Experimental dataset

Data	Values
Number of assembly workstations	9
Number of customer orders	50, 100, 200
Number of jobs/products per order	5
Number of job/product families	20
Standard processing time of each job/product	$U[60, 80]/\text{mins}$
Standard setup time for changeover	$U[5, 40]/\text{mins}$
Due time of customer orders	$U(0.6 \times D, 1.4 \times D)$ $D = \left(\frac{0.8 \times \textit{Total processing time}}{\textit{Number of assembly stations}}\right)$

Table 4 Uncertainties generation

Data	Values
Stochastic processing/setup time	
Variation coefficient	$cv = 0.3$
Standard job processing/setup time	p/mins
Actual job processing/setup time	$U\left[(1 - cv)p, (1 + cv)p\right]/\text{mins}$
New customer order arrivals	
Number of new customer orders	$1/2 \times$ *Total number of jobs*
Mean time between arrivals	$MTBA = 30/\text{mins}$
Time between order arrivals	$Exp(MTBA)$
Equipment failures	
Mean time between failures	$MTBF = 1000/\text{mins}$
Time between failures	$Exp(MTBF)$
Mean time to repair	$MTTR = 50/mins$
Time to repair	$Exp(MTTR)$

Table 5 Methods adopted as references in the performance evaluation

Methods	Description	Type	Real-time priority
MEDD	Modified earliest due date (Kanet and Li 2004)	Dynamic	$P = \max(pt_j, dt_j - \tau)$
JMEDD	Job with similar setup and modified earliest due date (Sharma and Jain 2015)	Dynamic	$P_1 = st_{j(s,t-1),j}$ $P_2 = \max(pt_j, dt_j - \tau)$
JIT	Just-in-Time (Đurasević and Jakobović 2018)	Dynamic	$P = (dt_j - pt_j - \tau)^2$

planning decisions (release to the shop floor) rely solely on P_2, whereas scheduling decisions (assignment to a workstation) depend on both P_1 and P_2.

4.4 *Performance Evaluation*

The numerical study utilizes the performance evaluation framework developed by Guo et al. (2021), encompassing cost-efficiency, punctuality, and simultaneity. For cost-efficiency, makespan (MS) and total setup time (TST) are standard metrics. Punctuality is assessed using mean order tardiness (MOTD) and the number of tardy orders (NTO). Simultaneity is evaluated through mean job flow time (MJFT) and mean order flow time (MOFT).

Tables 6 and 7 present the numerical outcomes for cost-efficiency, punctuality, and simultaneity across various methods in both deterministic and stochastic environments, with stochastic experiments conducted 200 times per setting. The findings reveal that HF-PSE consistently delivers superior performance in both environments, regardless of customer order volume. JMEDD generally ranks as the second most effective method. MEDD and JIT yield comparable results in cost-efficiency and punctuality, yet JIT excels in simultaneity. A notable trend is that the more real-time data incorporated into decision-making, the better the overall performance, warranting further investigation.

In the stochastic environment, 200 results are summarized by the boxplots in Fig. 5. It is evident that the proposed HF-PSE not only surpasses others on average but also exhibits less variability in statistics, particularly regarding TST, MOTD, MJFT, and MOFT metrics. This indicates that HF-PSE demonstrates superior performance stability and resilience compared to other methods in stochastic manufacturing settings.

Table 6 Performance of different approaches in deterministic environment

		50 orders				100 orders				200 orders			
		HF-PSE	MEDD	JMEDD	JIT	HF-PSE	MEDD	JMEDD	JIT	HF-PSE	MEDD	JMEDD	JIT
Cost-efficiency	MS	**2214.0**	2576.0	2255.0	2535.0	**4377.0**	5154.0	4499.0	5114.0	**8539.0**	10135.0	8695.0	10113.0
	TST	**2060.0**	5337.0	2399.0	5050.0	**3721.0**	10872.0	4712.0	10586.0	**6494.0**	20773.0	7806.0	20614.0
Punctuality	MOTD	**8.8**	126.3	16.4	107.1	**0.9**	393.1	22.5	263.9	**0.0**	574.1	2.0	354.4
	NTO	**6.0**	16.0	7.0	15.0	**4.0**	39.0	20.0	38.0	**0.0**	69.0	8.0	69.0
Simultaneity	MJFT	**299.7**	555.2	339.8	387.2	**389.1**	1098.6	435.4	503.5	**394.5**	2314.9	436.9	508.7
	MOFT	**354.1**	638.8	439.4	507.7	**445.2**	1710.4	538.1	753.0	**450.5**	3018.9	537.6	776.4

Table 7 Performance of different approaches in stochastic environment (mean value of 200 repetitions)

		50 orders				100 orders				200 orders			
		HF-PSE	MEDD	JMEDD	JIT	HF-PSE	MEDD	JMEDD	JIT	HF-PSE	MEDD	JMEDD	JIT
Cost-efficiency	MS	**2322.6**	2697.4	2376.2	2702.8	**4506.1**	5448.1	4699.8	5445.3	**8956.3**	10613.9	9202.1	10609.4
	TST	**1746.4**	5084.6	2299.4	5119.7	**3597.4**	10836.3	4462.2	10887.2	**6503.8**	20120.7	8089.5	20143.0
Punctuality	MOTD	**21.6**	194.0	42.7	170.6	**15.1**	545.5	89.0	366.7	**17.9**	852.3	73.7	521.1
	NTO	**8.4**	19.0	11.7	19.2	**13.9**	43.9	29.7	44.0	**18.7**	79.1	33.6	79.0
Simultaneity	MJFT	**290.3**	559.2	328.0	388.2	**348.3**	962.0	398.7	507.6	**357.2**	1780.4	403.2	511.4
	MOFT	**356.6**	690.0	440.7	534.1	**415.5**	1704.2	539.4	702.2	**422.4**	3020.0	727.2	802.5

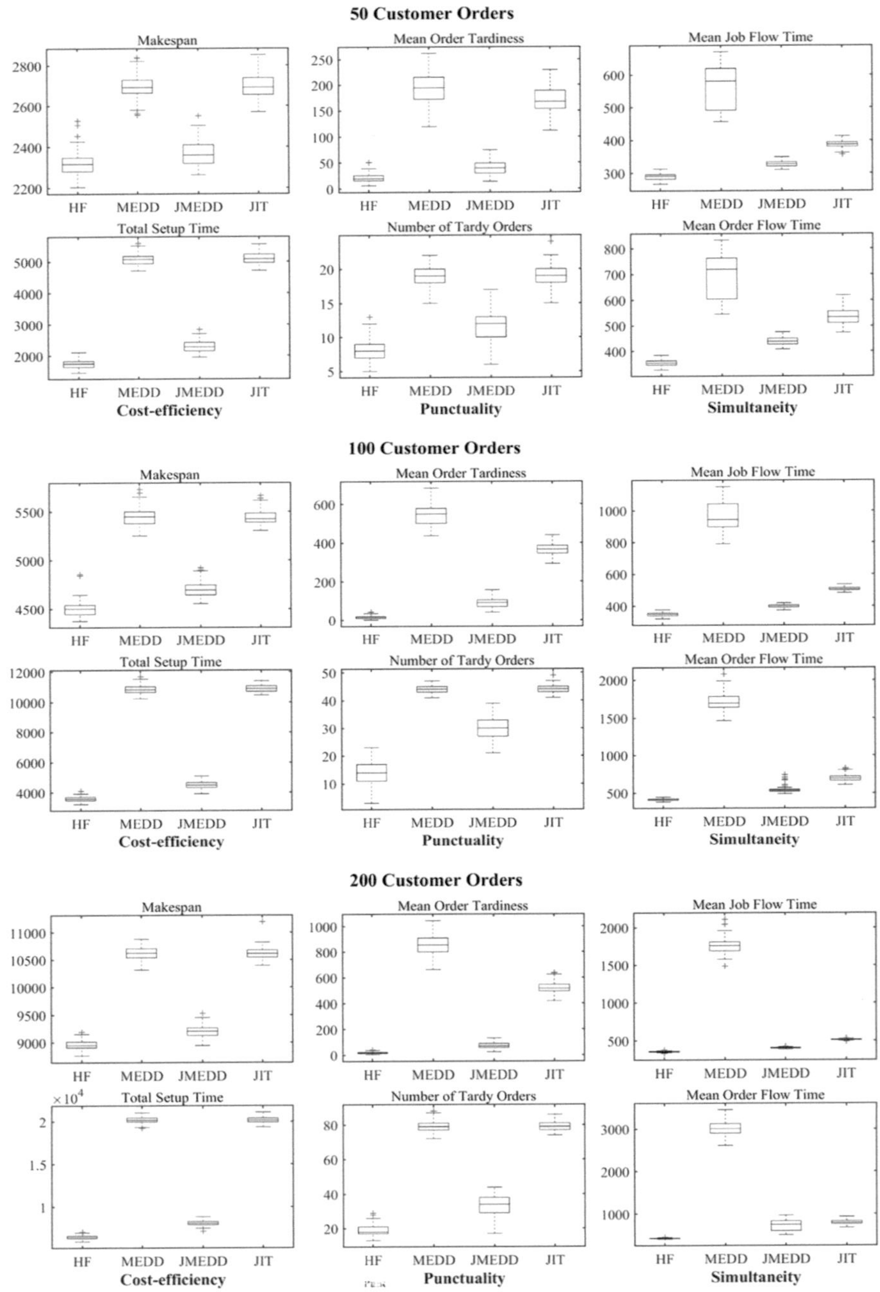

Fig. 5 Boxplots of performance measures

4.5 Discussions

The HF-PSE model offers three major advantages. First, experiments have confirmed HF-PSE's superiority in cost-efficiency, punctuality, and simultaneity, with stochastic settings further verifying its stability and resilience against typical manufacturing disruptions. Second, the model is exceptionally flexible in planning, scheduling, and execution. It can be easily tailored to fit other manufacturing systems' features and preferences. Third, HF-PSE creates a closed-loop information flow across planning, scheduling, execution, and feedback control by utilizing real-time data, bridging the gap between production plans/schedules and actual implementation.

From a practical standpoint, managers often prioritize the flexibility and efficacy of PSE tools (Tenhiälä 2011; de Man and Strandhagen 2018). HF-PSE offers a versatile production management solution that significantly enhances overall performance in terms of cost-efficiency, punctuality, and simultaneity. It also exhibits remarkable resilience and stability when addressing various uncertainties, which are crucial in industrial applications. Furthermore, as manufacturers embrace digital transformation in the Industry 4.0 era, vast amounts of shop floor data are gathered in real-time. HF-PSE integrates this real-time data into the decision-making process for further enhancements.

5 Conclusion

In conclusion, this research introduces an innovative and adaptable approach, Hyper-Factory Planning, Scheduling, and Execution (HF-PSE), tailored for production management in Industry 4.0 context. HF-PSE draws on the essential principles and flexibility found in university graduation ceremonies, featuring clustering-based planning with a rolling horizon to organize similar orders, real-time job scheduling according to workstation-job ticket suitability, and Out-of-Order ticketing to ensure seamless production flow. HF-PSE permits adaptability at every decision-making level to address manufacturing variability, avoiding the imposition of fixed processing sequences and timings. Furthermore, decisions are fully aligned with the real-time status of the shop floor, as the HF-PSE model integrates real-time data from Industry 4.0 settings to bolster overall resilience. Numerical analysis confirms the proposed HF-PSE's superiority.

The contributions of this study are encapsulated in three key aspects.

- It introduced an innovative PSE philosophy, inspired by graduation ceremonies, offering flexible and resilient principles to effectively bridge the gap between theory and practice in manufacturing systems.
- It formulated a comprehensive HF-PSE procedure, integrating real-time data into its mathematical models, thereby contributing to new theoretical research directions in production PSE for the Industry 4.0 era.

- Numerical experiments were conducted to validate the proposed method's effectiveness and resilience, revealing that HF-PSE performs better on average and demonstrates greater stability in stochastic environments.

In the future, the proposed HF-PSE could be expanded to incorporate intralogistics (material supply) into the overall framework and mathematical model, given the strong coupling between production and intralogistics. Furthermore, integrating big data-driven techniques with the real-time data-driven methods in HF-PSE could yield more effective and resilient solutions.

Acknowledgements This work was supported by the National Natural Science Foundation of China (52405548, 52375498), the Guangdong Basic and Applied Basic Research Foundation (2023A1515110622), the 2019 Guangdong Special Support Talent Program—Innovation and Entrepreneurship Leading Team (China) (2019BT02S593), and the Special Innovation Project of Guangdong Provincial Higher Education Universities (2024KTSCX009).

References

Allahverdi A, Ng CT, Cheng TE, Kovalyov MY (2008) A survey of scheduling problems with setup times or costs. Eur J Oper Res 187:985–1032
Benton W, Shin H (1998) Manufacturing planning and control: the evolution of MRP and JIT integration. Eur J Oper Res 110:411–440
Bi M, Kovalenko I, Tilbury DM et al (2024) Dynamic distributed decision-making for resilient resource reallocation in disrupted manufacturing systems. Int J Prod Res 62:1737–1757
Brah SA, Hunsucker JL (1991) Branch and bound algorithm for the flow shop with multiple processors. Eur J Oper Res 51:88–99
Chakhlevitch K, Cowling P (2008) Hyperheuristics: recent developments. In: Adaptive and multilevel metaheuristics, pp 3–29. Springer
de Man JC, Strandhagen JO (2018) Spreadsheet application still dominates enterprise resource planning and advanced planning systems. IFAC-PapersOnLine 51:1224–1229
Demirhan CD, Boukouvala F, Kim K, Song H, Tso WW, Floudas CA, Pistikopoulos EN (2020) An integrated data-driven modeling & global optimization approach for multi-period nonlinear production planning problems. Comput Chem Eng 141:107007
Duff IS, Koster J (2001) On algorithms for permuting large entries to the diagonal of a sparse matrix. SIAM J Matrix Anal Appl 22:973–996
Đurasević M, Jakobović D (2018) A survey of dispatching rules for the dynamic unrelated machines environment. Expert Syst Appl 113:555–569
Elyasi M, Altan B, Ekici A et al (2024) Production planning with flexible manufacturing systems under demand uncertainty. Int J Prod Res 62:157–170
Ghaleb M, Zolfagharinia H, Taghipour S (2020) Real-time production scheduling in the Industry-4.0 context: addressing uncertainties in job arrivals and machine breakdowns. Comput & Oper Res 123:105031
Guo D, Zhong RY, Lin P, Lyu Z, Rong Y, Huang GQ (2020) Digital twin-enabled graduation intelligent manufacturing system for fixed-position assembly islands. Robot Comput-Integr Manuf 63:101917
Guo D, Li M, Lyu Z, Kang K, Wu W, Zhong RY, Huang GQ (2021) Synchroperation in Industry 4.0 manufacturing. Int J Prod Econ 108171
Hazır Ö, Dolgui A (2013) Assembly line balancing under uncertainty: robust optimization models and exact solution method. Comput Ind Eng 65:261–267

Hubbs CD, Li C, Sahinidis NV, Grossmann IE, Wassick JM (2020) A deep reinforcement learning approach for chemical production scheduling. Comput Chem Eng 141:106982

Jiang Y, Li M, Li M et al (2022) Digital twin-enabled real-time synchronization for planning, scheduling, and execution in precast on-site assembly. Autom Constr 141:104397

Kanet JJ, Li X (2004) A weighted modified due date rule for sequencing to minimize weighted tardiness. J Sched 7:261–276

Laslo Z, Golenko-Ginzburg D, Keren B (2008) Optimal booking of machines in a virtual job-shop with stochastic processing times to minimize total machine rental and job tardiness costs. Int J Prod Econ 111:812–821

Li M, Zhong RY, Qu T, Huang GQ (2022b) Spatial–temporal out-of-order execution for advanced planning and scheduling in cyber-physical factories. J Intell Manuf 33:1355–1372

Li M, Huang GQ (2021) Production-intralogistics synchronization of industry 4.0 flexible assembly lines under graduation intelligent manufacturing system. Int J Prod Econ 241:108272

Li M, Ding H, Li M, Ling S, Huang GQ (2022a) Graduation-inspired synchronization for Industry 4.0 planning, scheduling, and execution. J Manuf Syst 64:94–106

Li M, Harish AR, Li M, Huang GQ (2023a) Blockchain-based fine-grained digital twin sharing framework for social manufacturing. Adv Eng Inform 58:102225

Li M, Guo D, Li M, Qu T, Huang GQ (2023b) Operation twins: production-intralogistics synchronisation in Industry 4.0. Int J Prod Res 61:5193–5211

Li M, Guo D, Li M, Qu T, Huang GQ (2023c) Real-time data-driven out-of-order synchronization for production and intralogistics in multiresource-constrained assembly systems. IEEE Trans Syst Man Cybern Syst

Li M, Cai Y, Guo D, Qu T, Huang GQ (2024) Data-driven diagnosis framework for platform product supply chains under disruptions. Int J Prod Res 1–23

Liu F, Wang S, Hong Y, Yue X (2017) On the robust and stable flowshop scheduling under stochastic and dynamic disruptions. IEEE Trans Eng Manage 64:539–553

Lu Z, Cui W, Han X (2015) Integrated production and preventive maintenance scheduling for a single machine with failure uncertainty. Comput Ind Eng 80:236–244

Magnanti TL, Vachani R (1990) A strong cutting plane algorithm for production scheduling with changeover costs. Oper Res 38:456–473

McFarlane D, Sarma S, Chirn JL, Wong C, Ashton K (2003) Auto ID systems and intelligent manufacturing control. Eng Appl Artif Intell 16:365–376

Mirabi M, Ghomi SF, Jolai F (2013) A two-stage hybrid flowshop scheduling problem in machine breakdown condition. J Intell Manuf 24:193–199

Mula J, Poler R, García-Sabater JP, Lario FC (2006) Models for production planning under uncertainty: a review. Int J Prod Econ 103:271–285

Nascimento MC, De Carvalho AC (2011) Spectral methods for graph clustering–a survey. Eur J Oper Res 211:221–231

Olsen TL, Tomlin B (2020) Industry 4.0: opportunities and challenges for operations management. Manuf & Serv Oper Manag 22:113–122

Oluyisola OE, Bhalla S, Sgarbossa F, Strandhagen JO (2021) Designing and developing smart production planning and control systems in the Industry 4.0 era: a methodology and case study. J Intell Manuf 1–22

Pinedo M (2005) Planning and scheduling in manufacturing and services. Springer

Pinedo ML (2009) Planning and scheduling in manufacturing and services. Springer Science & Business Media

Pinedo ML (2016) Scheduling: theory, algorithms, and systems. Springer

Qian C, Zhang Y, Jiang C, Pan S, Rong Y (2020) A real-time data-driven collaborative mechanism in fixed-position assembly systems for smart manufacturing. Robot Comput-Integr Manuf 61:101841

Sharma P, Jain A (2015) Performance analysis of dispatching rules in a stochastic dynamic job shop manufacturing system with sequence-dependent setup times: simulation approach. CIRP J Manuf Sci Technol 10:110–119

Tao F, Qi Q, Liu A, Kusiak A (2018) Data-driven smart manufacturing. J Manuf Syst 48:157–169
Tenhiälä A (2011) Contingency theory of capacity planning: the link between process types and planning methods. J Oper Manag 29:65–77
Zandieh M, Khatami A, Rahmati SHA (2017) Flexible job shop scheduling under condition-based maintenance: improved version of imperialist competitive algorithm. Appl Soft Comput 58:449–464
Zhong RY, Dai Q, Qu T, Hu G, Huang GQ (2013) RFID-enabled real-time manufacturing execution system for mass-customization production. Robot Comput-Integr Manuf 29:283–292
Zhou J, Li P, Zhou Y, Wang B, Zang J, Meng L (2018) Toward new-generation intelligent manufacturing. Engineering 4:11–20

Reconfigurable Manufacturing Systems for Industry 4.0: Concepts, Challenges, and Applications

Muhammad Ameer and Mohammed Dahane

Abstract This chapter provides an overview of how manufacturing systems have progressively evolved toward reconfigurable manufacturing systems (RMS) within the context of Industry 4.0. We focus on the key challenges in RMS design, the selection of reconfigurable system components, and process planning strategies in RMS environments. First, we offer a concise account of the historical shifts in production paradigms, highlighting the main drivers behind each transition. We then discuss the four industrial revolutions, setting the stage for next-generation manufacturing systems (NGMS). Among them, RMS is introduced with its fundamental principles and historical background. After presenting the six core characteristics of RMS that guide its design goals, we describe the various hardware components that can be reconfigured in an RMS. We then turn our attention to the evolution of computer-aided process planning (CAPP) in the RMS setting. Finally, we conclude by summarizing the essential insights related to RMS under Industry 4.0.

1 Evolution of Production Systems

Manufacturing systems are constantly evolving due to the uncertain nature of the market and technological advancements. Over time, social, economic, and technological developments and inventions are inevitable, leading to paradigm shifts in manufacturing systems. The first significant shift occurred approximately 10,000 years ago, marking the transition from foraging to farming, known as the Agrarian Revolution. During this period, humans utilized animal labor for activities such as production, transportation, and communication Schwab (2017), resulting in improved food production. Consequently, the human population grew, leading to the need for settlements and the emergence of cities. The 18th century marked the transition from the

M. Ameer · M. Dahane (✉)
Université de Lorraine, LGIPM-Metz F-57000, France
e-mail: mohammed.dahane@univ-lorraine.fr

M. Ameer
e-mail: muhammad.ameer@univ-lorraine.fr

S. Huang et al. (eds.), *Design and Operation of Smart Reconfigurable Manufacturing Systems in Industry 4.0/5.0*, Springer Series in Advanced Manufacturing,
https://doi.org/10.1007/978-3-032-00284-6_9

agrarian age to the industrial production age. In this chapter, we identify boundaries for classifying shifts in industrial systems, from existing manufacturing paradigms to new ones. Koren (2010) identified the boundaries between manufacturing revolutions in terms of the following paradigm shifts:

- Craft Production
- Mass Production
- Mass Customization
- Personalized Production.

As shown in Table 1, the first transition from the agrarian age to the industrial era took place in the 18th century, followed by a series of industrial paradigm shifts. The period and complexity of early manufacturing shifts are considerably greater than today. The spindle (the invention of the first industrial revolution) took almost 120 years to spread across Europe. Electricity, the foundation of the second industrial revolution, is still unavailable to 17% of the world's population. Similarly, more than half of the world's population lacks access to the internet. However, the internet spread exponentially faster than the spindle and electricity, reaching global adoption in less than a decade (Schwab 2017).

The drivers identified for production system shifts in the past were new scientific inventions, technologies, and the personalized needs of customers. For example, the invention of the Bessemer process enabled the mass production of steel in the latter half of the 19th century, paving the way for new technological advancements in the construction industry, such as the construction of skyscrapers. In contrast, present production system shifts are driven by uncertain markets and social needs. Market uncertainty can arise from economic factors, such as the transition from an industrial economy to a knowledge economy (Dźwigoł 2021), political uncertainty (Pertuze et al. 2019), environmental uncertainty (Aprisma and Sudaryati 2020), and innovation uncertainty (Jalonen 2012). Two of the predicted enabling drivers for future production system shifts are the products themselves and personalized needs within ecosystem constraints.

Production process history dates back to prehistoric times. Historians sometimes classify human civilization eras according to their production technologies, such as the Stone Age and the Bronze Age. Production techniques like casting, forging, and grinding date back to 4000 B.C. People who specialized in specific processes were named according to their crafts, such as blacksmiths who specialized in forging. These craftsmen were not limited to just the process but also handled all related business activities, including selling products directly to customers.

Koren (2010) represented global manufacturing revolutions in a product volume versus product variety diagram to illustrate the globalization event at the end of the 20th century. The classification of industrial paradigm shifts into four industrial revolutions is as follows:

Table 1 General history of production paradigm transitions

Production revolution	Time span	Shift	Source	Reason of Shift	Goal	Results
Agrarian revolution	10,000 BC	Foraging to farming	Muscle power	Utilization of animal and human effort	Production Transportation Communications	Improved food production Population growth Urbanization and growth of cities
Industrial transformation						
First industrial revolution	1760–1840	Farming to industry (craft production)	Mechanization Steam power	Use of machines for production Along with human craftsmanship	Industries Customer needs Scientific technologies	Railroad construction Steam engine invention Weaving loom industry
Second industrial revolution	Late 19th century	Craft production to mass production	Electric power	Electric power replaced mechanical power	Assembly lines Standardization	Numeric control machines Automobile industry Dedicated production lines

(continued)

Table 1 (continued)

Production revolution	Time span	Shift	Source	Reason of Shift	Goal	Results
Third industrial revolution (digital revolution)	Early 20th century	Mass production to mass customization	Semiconductor	Shift from human control to machine control (Digitalization) Machines became intelligent	Personal computer Internet CNC	Flexible manufacturing system Globalization Integration of world knowledge and market
Fourth industrial revolution	Start of 21st century	Customized and personalized production	Internet	Entire system becomes intelligent with machines	Smart factories Integration of all value chain cycle components	IoT Industrial digitalization CPPS RMS Intelligent products Integrated manufacturing systems Blockchain Connectivity

- **First Industrial Revolution**

In the second half of the 18th century, the invention of the steam engine marked the beginning of the industrial age. This period marked the transition from muscle power to mechanical power. Steam engines and coal replaced previously used sources like muscle power, water wheels, and windmills for powering factories. The invention of general-purpose machines occurred during this period. In Britain, during the 1760s, weaving loom machines were invented, revolutionizing the textile industry. The production and transportation sectors benefited the most during this period, with projects like railroad construction and the invention of the steam engine. The business models of that period were pull systems, where customers had to order products before production. For local markets, the production of customized products became possible. Production systems were highly dependent on skilled laborers or craftsmen and could produce a large variety of products but with minimal volume.

- **Second Industrial Revolution**

With the maturity of steam engines in almost every field in the latter half of the 19th century, inventions like the internal combustion engine and electricity unfolded the second industrial revolution. Steam engines were replaced by internal combustion engines. In production systems, machine power shifted from mechanical sources to electricity, leading to the invention of new products like the telegraph and the telephone, which revolutionized communication systems. The introduction of product standardization led to the assembly line concept, making the manufacturing of automobiles and airplanes possible. These inventions changed the dynamics of transportation systems. This period was known as the age of science and mass production. Due to the mass production paradigm, push economic models were implemented. The introduction of dedicated manufacturing systems (DMS) enabled the mass production of products.

- **Third Industrial Revolution**

The second half of the 20th century marked the third industrial revolution. The invention of integrated circuits and microprocessors following the transistor was critical for the field of electronics. This ultimately led to the spread of digitalization and automation in industrial sectors, resulting in automatic machines like computer numeric control (CNC) machines. The third industrial revolution is also called the computer or digital revolution. In the 1970s, the manufacturing of personal computers began. Production systems started shifting from mass production to mass customization during this period. The invention of flexible manufacturing systems (FMS) enabled the production of very complex products in low volume with high variety, such as ships and airplanes. The invention of the internet in the 1990s led to the digitization of the service industry. With the spread of high-capacity optic fiber across the ocean in the 1990s, connectivity changed from local to global. Alongside these technological advancements, significant political and economic changes occurred worldwide, as shown in Table 2 and represented on the map in Fig. 1 (Koren 2010). These factors made the globalization of production industries possible.

- **Fourth Industrial Revolution**

The fourth industrial revolution was conceptualized in 2011 in Vancouver, Germany, by a group of scientists in a project on a high-tech strategy for promoting computerization in the manufacturing sector. In 2015, Klaus Schwab coined the term Industry 4.0. The primary motivation of Industry 4.0 is the connectivity and integration of

Table 2 Significant events toward globalization (Koren 2010)

1	1991	Economic Liberalization Package by India
2	1992	Creation of the European Union
3	1992	Russia's Prices Were Freed, and President Yeltsin Started Enterprise Privatization
	1993	Boeing Design Center Established in Moscow
4	1994	*NAFTA (North American Free Trade Agreement—US, Canada, and Mexico)*
	1994	*GM Engine Part Manufacturing in China*
	1995	*Ford India Established (Joint Venture with Mahindra for Assembly of Ford Escort)*
	1995	*Delphi (Battery Production) Opens in China*
	1997	*General Motors Shanghai (GMS) in China*
	1998	*Daimler-Chrysler Formed by Merger of Daimler-Benz (Germany) and Chrysler Corp. (USA)*
5 a–h	2001	China Joined the World Trade Organization
	1992–2001	High-Capacity Transoceanic Optic-Fiber Cable Deployment Around the World (see Fig. 1)

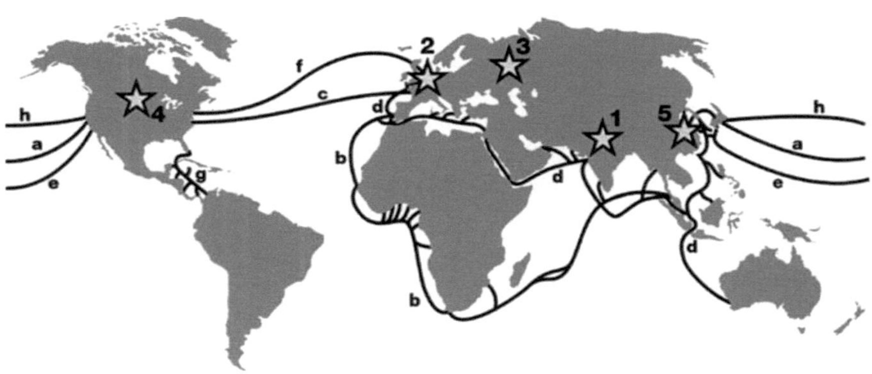

Fig. 1 Events across the world (Koren 2010)

manufacturing and service systems to provide effectiveness, adaptability, cooperation, coordination, and efficiency, as realized from design principles. After many advancements in Industry 4.0 over the past decade, the German platform Industrie 4.0 presented a 2030 vision titled "*Shaping Digital Ecosystems Globally*". A roadmap for the current decade encapsulates the trilogy of autonomy, interoperability, and sustainability as three development pillars toward a digital, networked, and sustainable manufacturing industry of the future (Sautter 2021).

2 Industry 4.0

According to (Mubarok 2020), Industry 4.0 is defined as "*a seamless collaboration of diverse advanced technologies that integrate smart visualization and analytics of autonomous processes to realize the vision of smart and intelligent factories, producing smart connected products and delivering smart services through realtime information transparency.*"

Figure 2 classifies 12 technologies as enablers for Industry 4.0. In 2015, the Boston Consulting Group (Rüßmann et al. 2015) identified nine technologies: IoT (Internet of Things), cloud computing, cybersecurity, big data and analytics, horizontal and vertical system integration, advanced robotics, additive manufacturing, augmented reality, and simulation. In 2020, (Mubarok 2020) added three more technologies: knowledge graphs, blockchain, and digital twins.

The beginning of the 21st century marks the onset of the Fourth Industrial Revolution, driven by exponential growth in modern technologies, globalization (facilitating market access), and evolving customer preferences and purchasing power. The evolution of existing production systems is imminent toward more intelligent and smarter production systems that can meet today's product demand and variances in an uncertain market. To explain the evolution of production systems in the era of Industry 4.0, (Morrar et al. 2017) developed a framework for the sustainability of evolving production systems, as shown in Fig. 3. The framework is based on three pillars: social, economic, and environmental. While providing a sustainable solution, the key drivers are social and technological innovations, which play a reciprocal role in the framework and act as an assessment mechanism for novel developments in Industry 4.0. The critical factor defined in the framework is designing a creative platform where experts from all related fields can work together toward novel solutions by using technological developments to make these existing systems sustainable.

3 Existing Manufacturing Systems

The introduction of new manufacturing technologies is influenced by changing product requirements driven by socio-economic needs. The mass production paradigm is associated with Henry Ford's invention of moving assembly lines, which

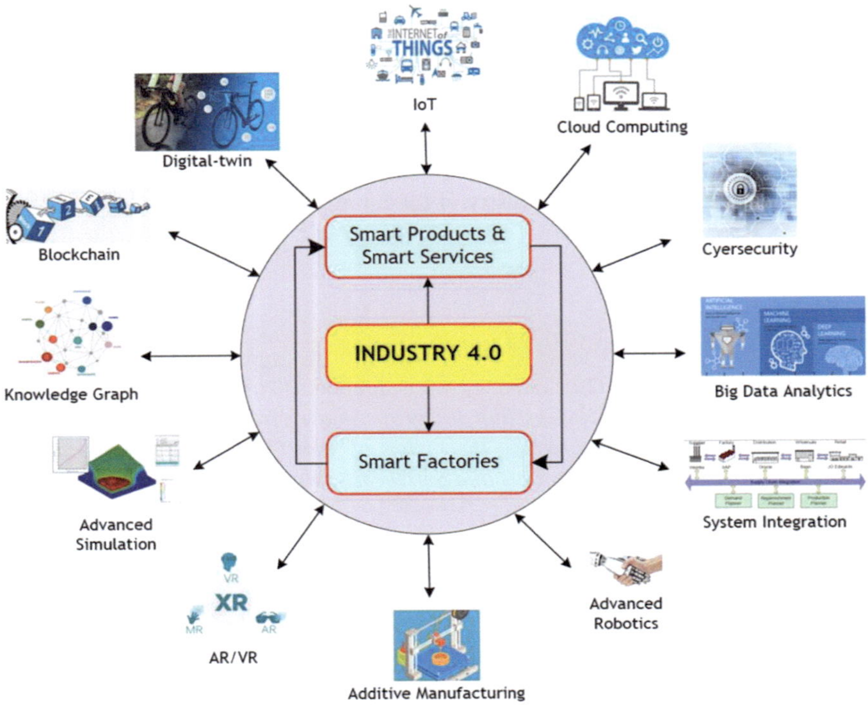

Fig. 2 Industry 4.0 enabling technologies by Mubarok (2020)

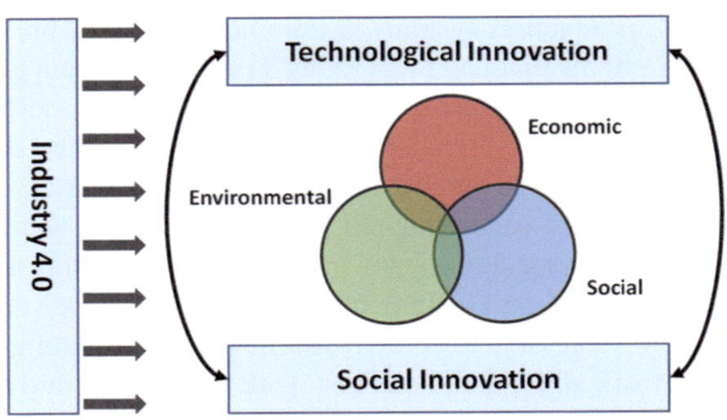

Fig. 3 Socio-techno-economic theory for sustainable Industry 4.0 by Morrar et al. (2017)

addressed cost-effectiveness. Similarly, Toyota began developing lean production principles (just-in-time) to address quality issues in their assembly line. At the same time, the development of CNC led to flexible manufacturing systems (FMS), which were developed to manage product variety (Koren et al. 2018). There are four basic ways to structure the processing space of a manufacturing system: (1) the job shop,

(2) the project shop, (3) the cellular system, and (4) the flow line. The first two are types of FMS; in the jobshop, machines are stationary, and the product is moved around, while in the project shop, the product (e.g., large ships or airplanes) is static, and machines are moved around the space. In manufacturing systems organized according to the cellular arrangement, equipment or machinery is grouped according to the process combinations that occur in families of parts (Khanna et al. 2014). Each cell contains machines that can produce a specific family of parts. In contrast, flow or transfer lines are dedicated manufacturing lines (DML) that can produce a single product (mostly) with high volume and quality. Conventional manufacturing systems are dominated mainly by flexible and dedicated manufacturing systems. Both have advantages: FMS can produce various products due to the usage of CNC machines with general flexibility, and DML has a high production rate and better quality. Today's uncertain market requires manufacturing systems to respond more effectively to change and meet product demand. Due to these perspectives, the system is required to handle high volume as well as variety. FMS and DML technologies are not designed to meet such demands. Therefore, next-generation manufacturing technologies (NGMS) have been introduced.

4 Reconfigurable Manufacturing Systems

In the late1990s, intense globalization revealed that existing production systems—namely, flexible manufacturing systems (FMS) and dedicated manufacturing systems (DMS)—were insufficiently responsive to evolving market needs and requirements. To address these challenges, a responsive, changeable, and reconfigurable manufacturing system (RMS) was developed. Koren et al. (1999) defined RMS as "*a system designed for rapid change in its structure, as well as its hardware and software components, to quickly adjust its production capacity and functionality in response to sudden market changes or intrinsic system changes.*"

4.1 Brief History

In the first decade of the 21st century, the introduction of RMS in the production of complex parts and systems—particularly engines, powertrains for cars, and trucks in the US automobile industry—significantly impacted the sector. Experts believe this was one of the significant factors in the recovery of the US automobile industry (Koren 2020). In the automobile sector, specifically the powertrain sector, cylinder heads, engine blocks, and gears were manufactured on transfer lines (dedicated production lines, DPL) until the introduction of RMS. These sectors benefited from RMS due to its responsiveness toward changes, providing the capability to address new market demand challenges.

In the second decade of the 21st century, the importance of RMS increased because of the enabling technologies of Industry 4.0, as described in the above section. These technologies are acting as powerful drivers of RMS. While enabling technologies of Industry 4.0 are transforming production systems to develop smart factories (James 2012), smart inventory management systems (Liang 2013), smart scheduling (Rossit et al. 2019), and smart supply chain management (Shao et al. 2021; Moeuf et al. 2017), production systems are adapting RMS at different levels, i.e., machine level, material handling level, layout level, and process planning level, because of RMS's ability to address the challenges of uncertain markets.

4.2 Characteristics of Reconfigurable Manufacturing Systems

To achieve the required design goals (low cost, high quality, and rapid responsiveness) of RMS, six core characteristics are defined (Koren 2013): scalability, convertibility, diagnosability, customization, integrability, and modularity, as shown in Fig. 4. These six core characteristics are widely used in the literature, either alone or in different combinations, to achieve the desired design goals. These characteristics are defined with their areas of application as follows:

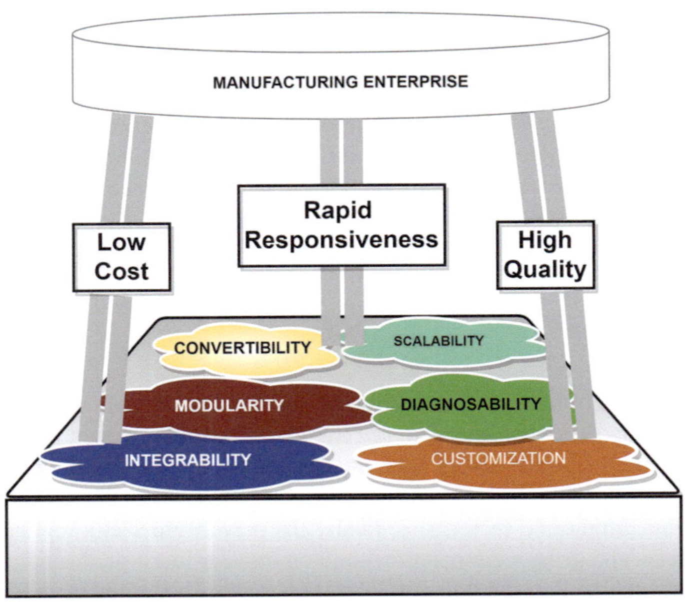

Fig. 4 RMS design goals and core characteristics relationship (Koren 2013)

- *Scalability*: The ability to design the RMS structure for adjustable capacity to deal with uncertain market demand. Structural adjustment involves adding or removing machines from the system layout at the system level, while at the machine level, it involves adding or removing modules of a machine (in the case of modular machines).
- *Convertibility*: The characteristic of RMS used for adjusting the functionality of the system and machines so that the system can be easily transformed for the production of new parts/products of the same family. Machines, configurations, and material handling systems contribute to rapid system conversion.
- *Diagnosability*: The ability of RMS to self-read the current state to detect and diagnose the root cause of part/product defects or machine failures and correct them accordingly (Bortolini et al. 2018), to reduce ramp-up time. In-line inspection machines' root-cause analysis in real-time during production is used for diagnosability. Due to its frequent reconfiguration, frequent diagnosability becomes crucial to avoid unnecessary breakdowns and maintain part/product quality.
- *Customization*: RMS is designed for a specific part or product family, so the flexibility of the system and machines is customized for the production of that part/product family. The aim is to reduce reconfiguration costs (Koren 2013).
- *Integrability*: The ability to connect modules rapidly and precisely through a set of mechanical, informative, and control interfaces, facilitating integration and communication (Bortolini et al. 2018). This reduces reconfiguration cost and time (Koren 2013).
- *Modularity*: All major components of RMS (structural elements, axes, controls, software, and tooling) are modular (Koren 2006). This reduces reconfiguration time, making the system more responsive to change.

Customization, scalability, and convertibility are the necessary characteristics for RMS, while modularity, diagnosability, and integrability are the supporting characteristics (Koren 2006).

4.3 Components of RMS

RMS consists of reconfigurable software and hardware components. Bi et al. (2008) further classified hardware components into four reconfigurable components, as shown in Fig. 9:

- Reconfigurable Machining System
- Reconfigurable Fixtures
- Reconfigurable Material Handling System
- Reconfigurable Assembly System.

The general description of RMS components is given below.

1. **Reconfigurable Machining System**
   The design of reconfigurable machines (RMs) is based on two principles (Koren 2006). Principle (1): RMs can be reconfigured for customized functionality and capacity adjustment of the production system. Principle (2): They are designed around the common characteristics of the part family. Different types of reconfigurable machines are classified into three groups by (Yelles-Chaouche et al. 2021):

   - **Reconfigurable Machine Tools**
     RMTs are structures designed to perform a specific customized range of machining operations, with the capability of configuration change to adapt to new machining operation requirements cost-effectively (Dashchenko 2006). In the literature, different RMTs have been designed for various machining operations (milling, drilling, boring, etc.), either from scratch or by modifying existing machines. Singh et al. (2019) presented methodologies for converting conventional drilling, lathe, and milling machines into reconfigurable machines. A detailed description of RMTs is provided in Sect. 4.4.
   - **Reconfigurable Assembly Machines**
     Assembly machines are mostly used in the assembly and transfer lines of automobile assembly. Katz (2007) designed a reconfigurable assembly machine (RAM) for automotive radiators, for tube and fin assembly. As shown in Fig. 5, RAM is a modular structure consisting of two areas: the loose matrix area and the assembly area.

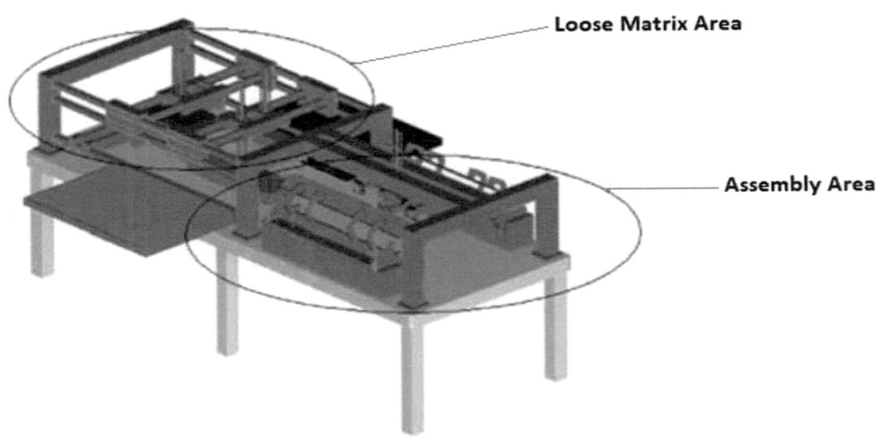

Fig. 5 Architecture of reconfigurable assembly machine adopted from (Katz 2007)

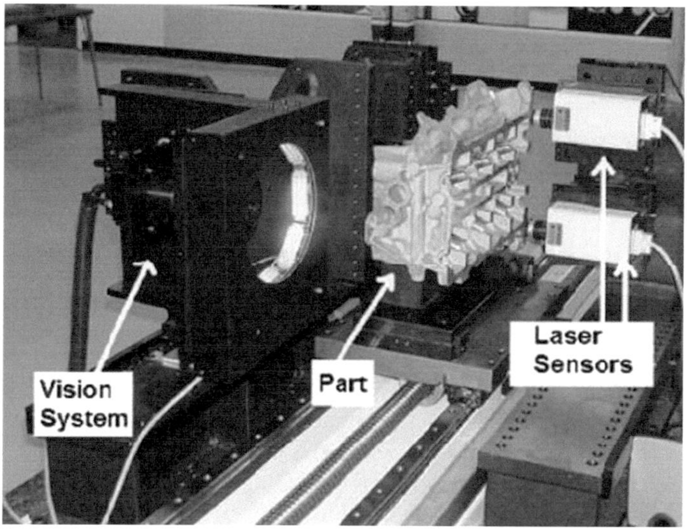

Fig. 6 Architecture of reconfigurable inspection machine adopted from (Katz 2007)

- **Reconfigurable Inspection Machines**
  The dynamic nature of RMS requires an inspection system that can constantly monitor the system's state (RMTs' performance changes with regular reconfigurations) along with product quality. Sautter (2021) developed a reconfigurable inspection system for monitoring product quality with the minimum number of reconfigurable inspection machines. Koren and Katz (2003) developed a patent for a reconfigurable inspection machine for inspecting cylinder head part families. As shown in Fig. 6, the RIM uses contact-less measuring methods based on commercial laser sensors and high-definition line-scan cameras in conjunction with computer vision technology and other commercially available non-contact sensors Katz (2007).

2. **Reconfigurable Fixtures**
   Fixtures are one of the significant cost drivers of the production system. Fixture design is for locating and holding the workpiece (product or part) accurately and precisely in a particular position repetitively for either machining or assembly operations. Therefore, they are mainly designed to be flexible to make the required adjustments for changing product demand. Jalonen (2012) presented different methods used to position and reconfigure flexible fixture devices. Asada and By (1985) used reconfigurable fixtures for kinematic analysis of part positioning in flexible assembly. The design of RMS is for a specific part/product family, so based on this concept, (Bejlegaard et al. 2018) designed a reconfigurable fixture capable of positioning the entire part family with the ability to efficiently convert for different part variants. As shown in Fig. 7, the reconfigurable fixture is adjusted for two-part variants swiftly.

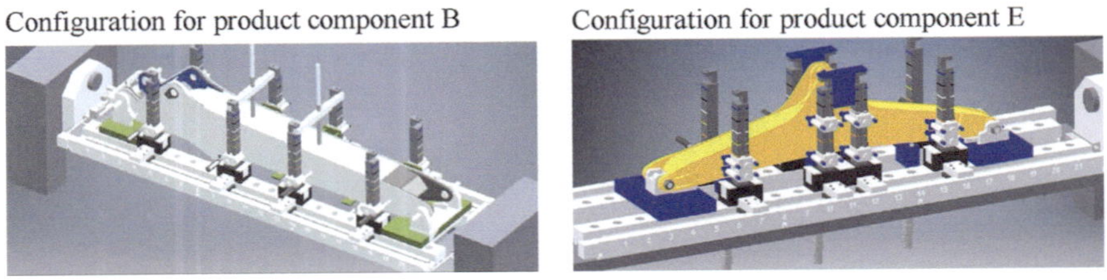

Fig. 7 Two part variants fixed in the same reconfigurable fixture in two different configurations. Figure adapted from (Bejlegaard et al. 2018)

3. **Reconfigurable Material Handling System**
RMS is a customized system in which different part variants are moved simultaneously to be processed at different workstations. Therefore, the material handling system is an integral part of RMS, whose main task is to provide reliable, accurate, and efficient transfer of part variants according to process scheduling, without bottlenecks and stoppages. In the literature, different material handling systems or components of the system have been developed. Koekemoer and Gorlach (2018) developed a reconfigurable pallet system to facilitate automated robotic assembly for a highly customized production environment. (Mohamad et al. 2018) proposed an integrated system architecture for reconfigurable conveyor systems (RCS) to support frequent changes in customer demand in a manufacturing system. As shown in Fig. 8, the RCS architecture consists of six modules of lab-scale conveyors. Each module has three outputs: sensors, actuators, and motor drives. Each module has an Omron CP1L—EL20DR-D controller, which provides programming control (modifying, removing, and adding programs) without stopping operations. The central system controller used for automation is a PC-based controller, while PLC controllers are used for sub-system control. The physical configuration has two conveyor types: straight-line and pneumatic cylinder conveyors. Four possible configurations can be produced from these conveyors: straight-lineshaped, U-shaped, L-shaped, and closed-loop-shaped.
4. **Reconfigurable Assembly System**
Reconfigurable assembly systems are based on the idea of a fundamental shift from product-specific conventional automation (flexible automation) to reconfigurable automation. Flexible production system automation is for a family of predefined products, while reconfigurable automation is for a dynamic mix of products flowing across the assembly system. Modularity and reconfigurability are the two main characteristics that separate the two. To design and implement reconfigurable automation for agile manufacturing systems, (Molnar et al. 2006) proposed a three-layer (base, process, and product) system architecture to develop standard reconfigurable assembly cells (Fig. 9).

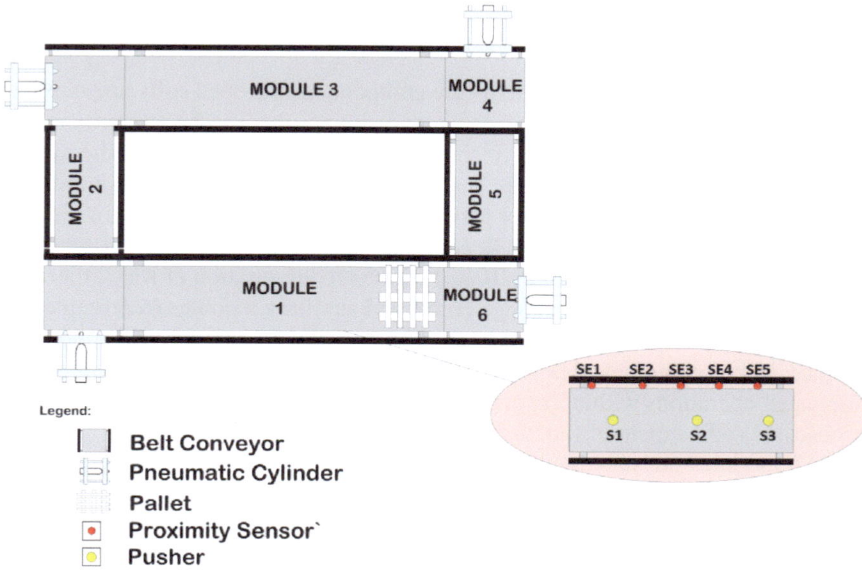

Fig. 8 Architecture of reconfigurable conveyor system adopted from (Mohamad et al. 2018)

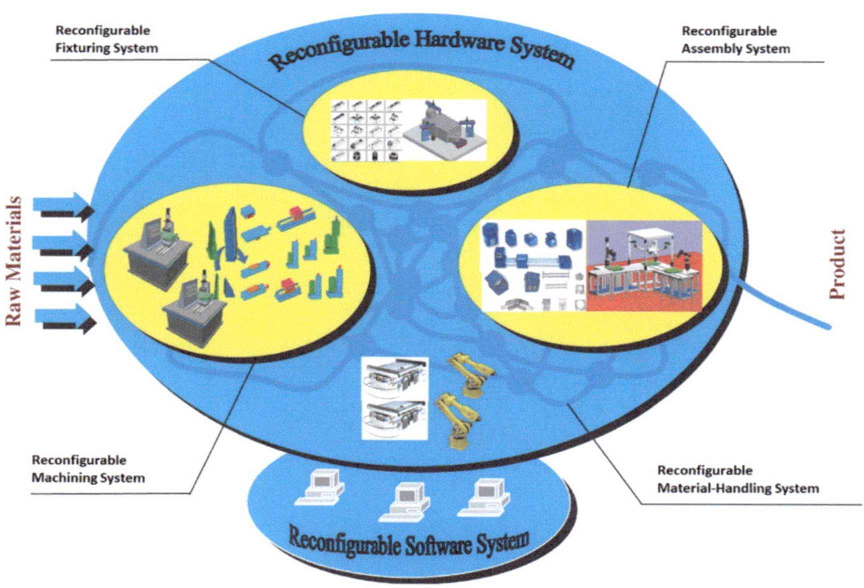

Fig. 9 RMS components by Bi et al. (2008)

4.4 Reconfigurable Machine Tools

RMT is a structure with modular hardware components and open software architecture (Kumar etal. 2021).RMTs are intelligent machines that can be promptly reconfigured for customized functionality and capacity adjustment, making the system responsive to unintended changes. The RMT is designed to perform necessary machining operations on all member parts of a given part family by reconfiguring the machine tool (Landers et al. 2001). Reconfiguration involves the addition, removal, or adjustment of different modules. Based on the reconfiguration of RMT, there are basic modules (represented by orange color) and auxiliary modules (represented by green color), as shown in Fig. 10. Basic modules of RMT are fixed, while auxiliary modules are reconfigurable. RMT can be reconfigured to specific configurations for anticipated operations within the part family. Reconfiguration is carried out for two purposes: (1) adjustment of functionality and (2) adjustment of capacity.

RMT modules are classified into three categories by Gadalla and Xue (2017) in their literature review of RMT: (1) processing modules, (2) platform modules, and (3) accessory modules. The module assembly mechanism uses three types of

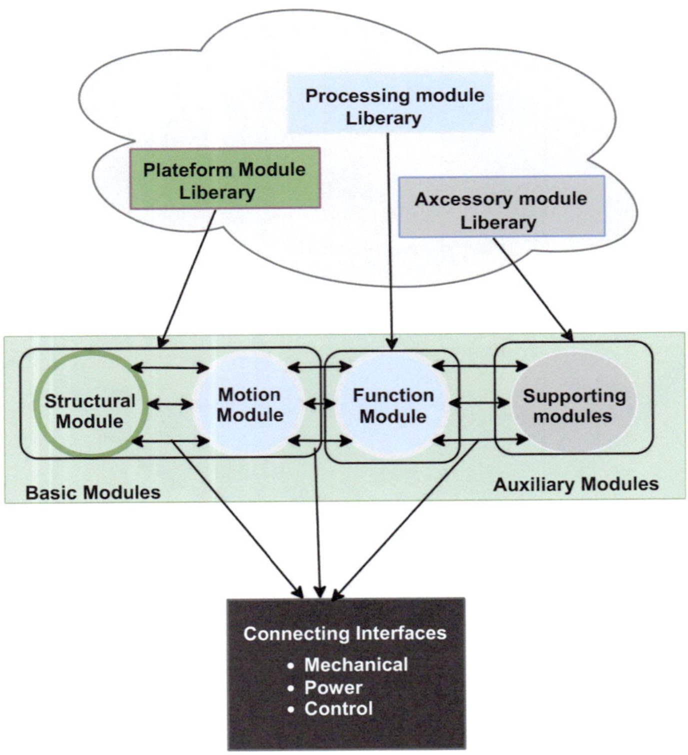

Fig. 10 Modular RMT general architecture

connection interfaces: mechanical interface, power interface, and control interface, as shown in Fig. 10. *Processing modules* are auxiliary modules that are interchangeable. Each processing module has specific functionality to perform different machining operations like drilling, milling, boring, etc. They are responsible for the functionality change of the RMT. This means the interchange of these modules as end effectors results in a change in the RMT's functionality. Thus, the configuration change of RMT for functionality adjustment is defined by processing modules. *Platform modules* consist of both basic and auxiliary modules. They form the main structure of the RMT. Platform modules can be classified into motion modules and structural modules. Motion modules are auxiliary modules controlling different axes of motion (degrees of freedom, DOF) of the RMT. The reconfiguration of motion modules provides a change in the degree of freedom to adjust to the required tool approach direction (TAD) of the RMT. This reconfiguration of motion modules is used to achieve the customized capacity adjustment of a given part family, making RMS capable of scalability. Three degrees of freedom (DOF) (in a Cartesian coordinate system) of RMT means the current configuration of motion modules provides three translatory motions along the X, Y, and Z axes. For scaling up the same RMT from three to four DOF, a rotational axis motion module can be added, changing the configuration of the RMT. The configuration of RMT for capacity adjustment is defined by motion modules, while platform modules are basic modules that shape the structure of the RMT. *Accessory modules* are secondary modules used for supporting purposes in machining operations, such as fixture modules for workpiece clamping.

The mechanical interface facilitates the physical interconnection between modules and the connection of different modules to a single interface, enabling the RMT topology (Padayachee et al. 2009). The power interface provides the necessary power for enabling different types of motions of the RMT. Each module has a standard control interface, which can collect sensor information and receive module commands with feedback channels.

Gadalla and Xue (2017) classified research in advancements of RMT design problems into three categories: architecture design, configuration design, and system integration and control. From an architectural perspective, they divided RMTs into two groups:

1. **Modular Architecture**
   Padayachee et al. (2009) designed a modular reconfigurable machine (MRM) capable of performing one operation in one configuration. Wang et al. (2019) developed a methodology to cluster module groups for designing modular RMT, enabling the modular RMT to perform more than one operation in one configuration. Modules can be added, removed, and adjusted to achieve different configurations of RMT for required changes. The design of open-architecture modular RMTs allows for the inclusion of third-party external modules by keeping open interfaces between modules. Modular RMTs are divided into two categories based on differences in their interfaces:

- Semi-open architecture interface
- Open architecture interface

2. **Integral Architecture**
   Too frequent reconfiguration of modular RMTs requires considerable effort. To reduce this effort, RMTs with integral architecture have been developed.

The design limitation of integral architecture RMTs restricts them to a specific family. Dhupia et al. (2007) developed an arch-type RMT for specific engine block families (V8 and V6), which has customized flexibility. The arch-type structure design helps fix the tool spindle in different positions to adjust the DOF of the RMT accordingly. Arch-type RMT is capable of performing one operation in one configuration. At the same time, (Son et al. 2010) designed a multi-arch-type RMT, which can perform more than one operation in the same configurational arrangement.

5 Process Planning

Process planning is a critical component of the manufacturing cycle, meticulously defining the process of transforming raw materials into a specific form. This form must be designed by the product designer and expressed in drawings, geometric dimensioning, and tolerance (GD&T) codes used by Halevi and Weill (1994). Process planning is a knowledge-intensive transformation activity between the design and manufacturing phases of the production system. By respecting precedence constraints, operations are sequenced. For each operation, a machine is selected, forming the layout configurations of the production system. Process planning activities involve process selection, operation sequencing, and the allocation of suitable machining equipment according to part design specifications and manufacturing conditions (Deja and Siemiatkowski 2013).

In manual process planning, reliance on expert process planners often leads to inconsistencies in handling complex, repetitive tasks, reducing system productivity and compromising product quality. On the other hand, computer-aided process planning (CAPP) uses computers to automate repetitive tasks, improving productivity.

The idea of CAPP stems from the need to automate the manual process plan.

CAPP bridges the computer-aided design (CAD) and computer-aided manufacturing (CAM) phases of the manufacturing system life cycle. The position of CAPP in the manufacturing cycle, along with ten technologies used for its development, is shown in Fig. 11, as identified by Yusof and Latif (2014). The design process of CAPP from an automation perspective has three categories: feature recognition from CAD, the decision-making phase, and the development of a source file. Kumar (2017) in their review related to the application of artificial intelligence in CAPP, classified the literature works of CAPP into three categories: feature-based design, expert systems, and evolutionary approaches. Feature-based designs are the primary inputs to CAPP, while expert-based and evolutionary approaches use artificial intelligence techniques for decision-making in CAPP. Wu et al. (2021), in their literature review of CAPP, identified advancements made by immersive visualization technologies in the field of CAPP and highlighted the challenges faced by CAPP in integrating CAD and CAM in product life cycle management (PLM). In the manufacturing cycle, the required information in the form of operations is interpreted from the given design (CAD) of the part/product, along with relevant data for machining these operations (in the form of different data models, such as feature-based models), to generate CAPP. The output of CAPP includes sequenced operations with selected machines for each operation, which serves as input to computer-aided manufacturing (CAM).

Besharati-Foumani et al. (2019) presented a literature review on intelligent process planning for manufacturing systems. The objective functions considered are cost and time minimization for increasing productivity and minimizing errors for quality improvement.

Khanetal (2022) presented a literature review of process planning in reconfigurable manufacturing systems, focusing on optimization efforts. Figure 12 shows their classification of process planning literature for RMS. The two-phase division of process planning literature includes the theoretical and implementation phases.

In the past two decades, designers have focused on the design and operation of manufacturing systems and product design to improve system productivity instead of automating manufacturing processes or technological advancements (ElMaraghy 2005). Today's market uncertainties propel manufacturing systems to be more dynamic and responsive toward simultaneously addressing product variety and quantity demand changes. The realization of system responsiveness at the design level is achieved through system integration. CAPP design is inherently deterministic and implemented for static manufacturing systems. Consequently, it cannot cope with product variety and dynamic manufacturing systems. Considering the dynamic nature of manufacturing systems, reconfigurable process planning (RPP) requires the generation of different performance indicators. These indicators are used alongside

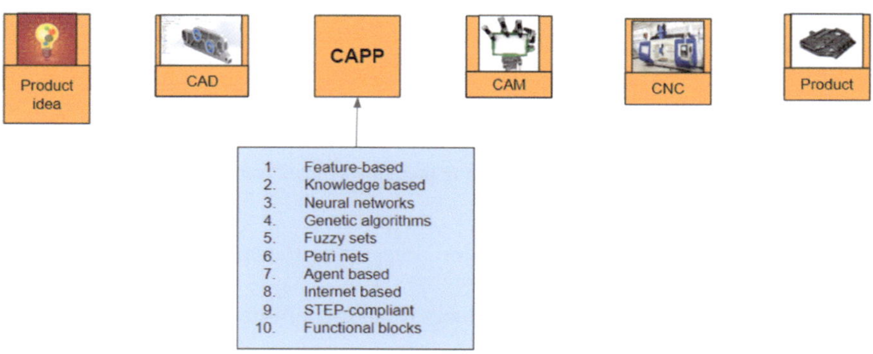

Fig. 11 CAPP position in a manufacturing system with ten development technologies by Yusof and Latif (2014)

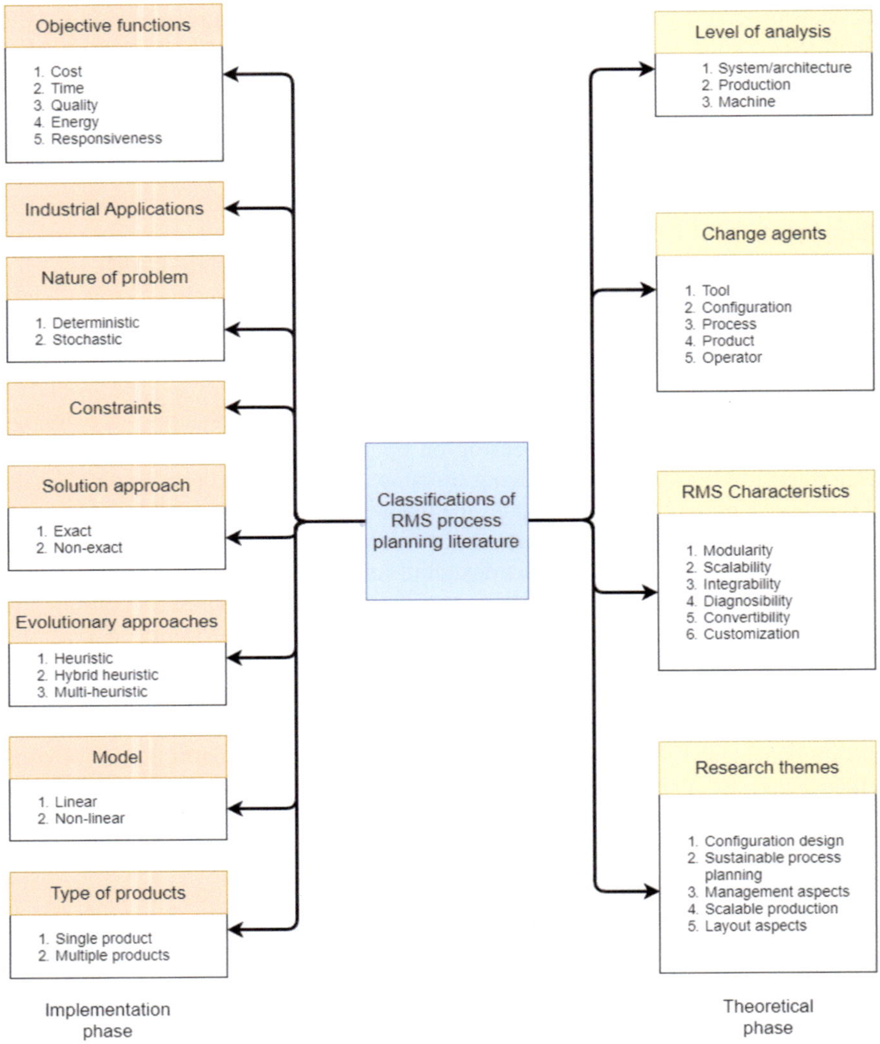

Fig. 12 Classification of RMS process planning literature adopted from Khan et al. (2022)

objective functions of process automation, integrating the entire manufacturing life cycle with automation.

This integration helps improve productivity in modern-day developing manufacturing systems.

The two phases of classification for the evolution of process planning are (1) the introduction of CAPP and (2) the introduction of RPP. In the first phase, the focus is on automating manual processing activities to achieve desired objective functions. In the second phase of RPP, the design focus shifts toward integrating the manufacturing system life cycle by considering different performance indicators to achieve the desired objective functions.

5.1 Computer-Aided Process Planning

Niebel (1965) introduced the concept of CAPP to increase productivity and quality by improving the speed and accuracy of manual process plans through the automation of as many manual tasks as possible (Yip-Hoi 2017). As most activities in computer-aided design (CAD) and computer-aided manufacturing (CAM) were already automated, process plan automation became the bottleneck problem of the manufacturing production cycle. The challenge recognized in the transition from manual process planning to CAPP is the dependency on experienced process planners in the decision-making process.

Over the past four decades, significant research has addressed the problem of minimizing dependencies by automating either a small portion of the process plan or the entire decision-making process. Yusof and Latif (2014) identified ten technologies used in CAPP for CAD and CAM integration and automation of the decision-making process in the production cycle, as shown in Fig. 11. These technologies include feature-based systems, knowledge-based systems, artificial neural networks, genetic algorithms (GAs), fuzzy set theory and fuzzy logic, Petri nets (PNs), agent-based technologies, Internet-based technologies, a standard for the exchange of product data (STEP)-compliant methods, and functional blocks. The benefits of CAPP over manual process planning, identified by Chryssolouris (2006), are: *improved productivity*, *low production cost*, *consistency*, *time savings*, and *rapid integration of new production capabilities*.

In the literature, the design process for CAPP has been divided into three parts: (1) feature recognition from CAD, (2) process plan design (decision-making phase), and (3) development of a detailed source file for CAM (Chryssolouris 2006). Feature recognition allows the CAD product design to be interpreted as input to CAPP, for which feature-based designs are being developed (Kumar 2017). Process plan design consists of different decision-making steps, including operation selection, sequencing, and assignment of machines and tools for each operation, explained in the next section. The data generated is then compiled into a standard source file used as input to CAM and as source code for CNC machines. Fundamentally, the objectives of process plan development are reducing throughput time and improving the quality of the part to be produced (Halevi and Weill 1994) (Fig. 13).

The core design of the process plan revolves around the decision-making process among available alternative options. There are three stages of selection between alternative options: operation selection, operation sequencing, and operation assignment. For each stage, there are different available alternatives. These decision making steps are complex due to the many parameters with alternative selection options. The design problem of CAPP has been acknowledged as NP-Hard, so two approaches

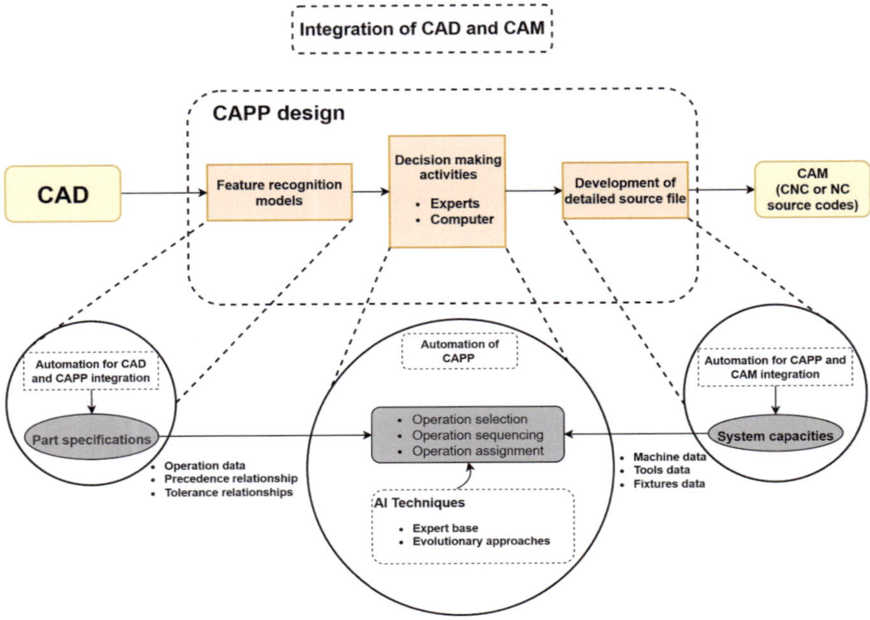

Fig. 13 CAPP in the manufacturing process (Yusof and Latif 2014)

used in the literature to solve this problem are (1) expert-based and (2) evolutionary approaches, reviewed in Wu et al. (2021) and Kumar (2017).

Both approaches use different types of artificial intelligence techniques for decision making, as detailed in Besharati-Foumani et al. (2019).

For improving system productivity, at the process design level, cost and time minimization problems are considered in the literature, as briefly discussed below:

1. **Cost**: The usual parameters in the cost minimization problem are machine cost, tool cost, setup cost, machine change cost, and tool change cost (Wu et al. 2021; Petrović et al. 2016). All these parameters are related to machines, cutting tools, and fixtures. Machine cost refers to the cost of machines that perform all machining operations in the given process plan. Tool cost is the cost of cutting tools used in the process plan. Setup change cost is the cost of total setup changes between two consecutive operations of the process plan. Machine change and tool change costs refer to the total machine and tool change costs between two operations of the given process plan.
2. **Time**: Common parameters for time include processing time, transportation time, total tool change time, and total setup change time Petrović et al. (2016). Like cost parameters, they are also related to machines, tools, and fixtures. One additional parameter is the material handling system (material handling between machines is mostly ignored at the process planning stage, so only fixtures (material handling on the machine) are considered).

3. **Quality**: Error minimization has been carried out for quality improvement. The quality of the part depends on system capabilities, such as the type of precision tools, fixtures, and machines. The manufacturing process controls the required tolerance limits of operations so that the part produced is within acceptable quality limits. In process planning, quality is addressed through two types of constraints between operations (Wu et al. 2021):
   - Precedence constraints are priority relationships between operations. They control operation sequencing in the process plan (the second step in decision-making activities) but do not directly affect part quality.
   - Tolerance constraints are geometric design constraints between operations. Every operation has design tolerance, which depends on the selection of the datum. Tolerance constraints are relationships between operations that control operation tolerance (defined by machining precision limits) to be within design tolerance limits. Handling tolerance constraints in the process plan controls manufacturing errors. Tolerance is further classified into local and global tolerance by Mei et al. (1995). Local tolerance is for a single operation, depending on the quality of the tool and the precision of the machine used for the operation. Global tolerance is between two or more operations in the process plan. The selection of the datum for operations and the setup of the part in the fixture during machining affect global tolerance.

The transition from manual process planning to CAPP has enabled the goals of increasing productivity and improving quality by automating the process plan to integrate CAD and CAM in the manufacturing life cycle. However, in the past two decades, productivity improvement has been achieved through the design and operation of manufacturing systems and product design, rather than through automation of the manufacturing process or technological improvements (ElMaraghy 2005). Along with significant production paradigm shifts due to the transformation of systems toward the Industry 4.0 paradigm, manufacturing paradigm shifts toward mass individualization (Gu and Koren 2022), mass customization Tseng and Jiao (2001), and co-evolution (Tolio et al. 2010) have given rise to product variety management concepts for uncertain markets. Considering product variety design as the main factor for productivity improvement requires the integration of the entire manufacturing system life cycle. This shifts the focus of process automation toward system integration. Therefore, next-generation manufacturing systems are designed with customized modularity, such as reconfigurable manufacturing systems (Koren et al. 1999). Consequently, considering only process plan automation is becoming an obsolete idea. The process plan must adapt to Industry 4.0 by considering product variety and regular system reconfiguration.

5.2 Reconfigurable Process Planning

Reconfigurable process planning (RPP) is tailored for reconfigurable manufacturing systems (RMS). RMS features a modular design and customized responsiveness to achieve scalability and manage part variety for specific part families. Mature and evolving part families are the two types of part families defined in RMS literature. Mature part families have fixed boundaries, meaning no external parts can be added or removed (Xia et al. 2018). In contrast, evolving part families have transparent boundaries, allowing new parts to be added or removed from the family (Azab and ElMaraghy. 2007). The responsiveness of RMS toward change is due to the different reconfigurable components of the production system. One system component is the modular machine, called a reconfigurable machine (RMT). RMTs have the capability of reconfiguration toward customized changes for a given part/product family. This induced reconfiguration of RMTs makes the system more responsive to uncertain market conditions, increasing the impetus on the decision-making step of process planning.

A reconfigurable process plan (RPP) is generated for a given part of the part family. Inputs to RPP are the design requirements of the part and the reconfigurable production system capabilities, as shown in Fig. 14. Market requirements assess the need for scalability of the existing part or the manufacturing of a new part of the same part family and provide feedback to the part design database and production system databases. The part design database generates a new product design by adding or removing features. In the production system, the required component selection is based on the feedback. The process plan is reconfigured according to the new part design, and for the new product, RPP selects the best machine (RMTs) with configurations and tools, as well as fixtures. This cyclic reconfiguration of the production system for part variety management can affect the productivity and quality of the system. For this purpose, performance indicators are defined in the literature (discussed in the section below). Therefore, in the design of RPP, along with automation objective functions, performance indicators are included, which help integrate the production system life cycle.

The decision-making steps of RPP design are the same as CAPP: operation selection, operation sequencing, and operation assignment. Since components of the production system, such as machines and fixtures, are modular and regularly reconfigured, their reconfiguration is also taken into account alongside production system capabilities (Ameer and Dahane 2020; Bensmaine et al. 2012; Azab and ElMaraghy. 2007; Baqai 2010). For the decision-making steps of RPP, four solution approaches are used: mathematical programming models, dynamic programming, heuristics, and meta-heuristics (Brahimi et al. 2019). These evolutionary approaches consider the objective functions of cost, time, and quality (as discussed in the CAPP section above). The parameters are related to machines, tools, and fixtures, along with consideration of the configuration of these system components. For the objective functions of cost and time, reconfiguration of machines and processing times are considered alongside other parameters. For the objective function of quality, an

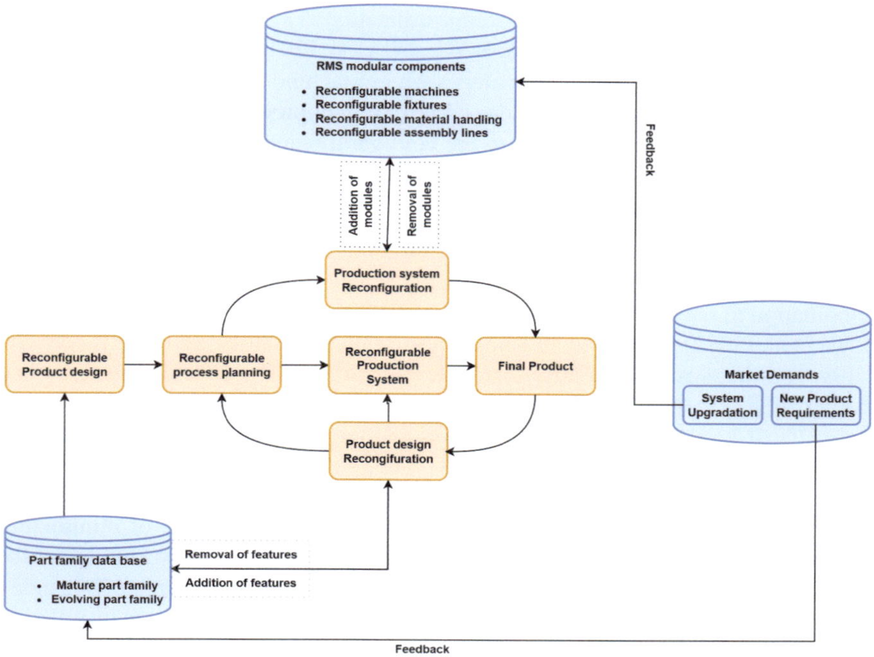

Fig. 14 RPP design toward integration of manufacturing system life cycle

additional parameter of reconfiguration effort is considered Bensmaine et al. (2012), along with process error reduction.

1. **Cost**: At the process planning level, the total cost is the sum of capital cost and operating cost (Yelles-Chaouche et al. 2021). Capital cost is the fixed, onetime cost of system components, such as RMTs, modules, spindles, tools, and fixtures. Operating cost includes usage and maintenance, energy consumption, and RMT reconfiguration costs of the system components (Ameer and Dahane 2020; Bensmaine et al. 2012).
2. **Time**: The total time considered in RPP consists of the sum of all processing times, including RMT processing time, tool usage time, RMT change time, tool change time (Bensmaine et al. 2012), fixture change time (part setup change time) (Ameer and Dahane 2020), transportation time, and RMT configuration change time (Goyal et al. 2012).
3. **Quality**: The parameters considered for maximizing quality are related to the type of constraints, as explained in the section above. The additional parameter considered due to the system's regular reconfigurability of RMTs to handle part variety is the reconfiguration effort, defined in (Prasad and Jayswal. 2018; Bensmaine et al. 2012; Goyal et al. 2012).
4. **Additional Parameters**: Along with cost, (Ashraf and Hasan. 2018) considered the reliability and operational capabilities of RMTs as objective functions.

Goyal and Jain (2016) defined machine utilization as a conflicting objective function with cost. Xia et al. (2018), Benderbal et al. (2015), Azab and ElMaraghy (2007) defined additional parameters such as machine precision, machine availability, and reconfigurability index. The introduction of all these parameters at the process level design is due to the configurational capabilities of RMTs for product variety management.

Azab and ElMaraghy (2007) developed a *reconfigurability index* (*RI*), which accounts for all changes made due to reconfiguration in the existing system of the process plan. The reconfigurability index developed in different works considers a combination of the six core characteristics of RMS ElMaraghy (2005). For example, Goyal et al. (2012) developed an RI by considering modularity, scalability, convertibility, and diagnosability. Farid (2017) developed an RI by considering three key enablers of RMS: integrability, convertibility, and customization. Dahane and Benyoucef (2016) developed an RI that considers handling errors due to the reconfiguration of RMTs after each part production.

Benderbal et al. (2018) developed a *modularity index*, which aims to maximize modularity in the given RMS design alongside objective functions of minimizing cost and time. Benderbal et al. (2015) defined a *flexibility index*, which looks for several possible alternatives to perform the operations of the process plan. This index provides an alternative solution if an RMT is unavailable for an operation by utilizing the system's modularity, making the system more responsive to change. Ma et al. (2020) developed a *process and feature commonality index*, used for decision making in operation sequencing of RPP for product variety management in open manufacturing. The developed index acts as a performance indicator of crowdsourcing in open manufacturing at the process planning level for managing product variety.

6 Conclusion

In a world marked by uncertain markets and rapid technological advancements, manufacturing systems must continuously adapt to remain competitive. This chapter explored the historical progression of manufacturing systems through four industrial revolutions, culminating in the Industry 4.0 era. We discussed the principal enabling technologies behind Industry 4.0 and the societal and economic shifts driving manufacturing transformations. To address new market and technological challenges, next-generation manufacturing systems such as RMS have emerged. We detailed RMS's six core characteristics—scalability, convertibility, diagnosability, customization, integrability, and modularity—and described how its various hardware components collectively enable rapid responsiveness while controlling costs and preserving quality. Furthermore, we examined the evolution of process planning, from manual methods to computer-aided process planning (CAPP), and subsequently to reconfigurable process planning (RPP) within RMS. This evolution highlights the necessity of integrating reconfiguration-based performance measures to maintain system agility and meet the demands of modern manufacturing. As industries continue to evolve, the integration of emerging technologies such as AI and IoT into RMS will be crucial for addressing future challenges and unlocking new opportunities for innovation and efficiency.

References

Ameer M, Dahane M (2020) New heuristic for setup plan generation and multi tool modules selection in reconfigurable manufacturing systems. In: 2020 7th International conference on control, decision and information technologies (CoDIT), vol 1, pp 1069–1074. IEEE

Aprisma R, Sudaryati E (2020) Environmental uncertainty and firm performance: the moderating role of corporate governance. Jurnal Akuntansi 24(2):187–203

Asada H, By A (1985) Kinematic analysis of workpart fixturing for flexible assembly with automatically reconfigurable fixtures. IEEE J Robot Autom 1(2):86–94. https://doi.org/10.1109/JRA.1985.1087007

Ashraf M, Hasan F (2018) Configuration selection for a reconfigurable manufacturing flow line involving part production with operation constraints. Int J Adv Manuf Technol 98(5):2137–2156

Azab A, ElMaraghy H (2007) Mathematical modeling for reconfigurable process planning. CIRP Ann 56(1):467–472

Baqai A (2010) Co-conception des processus d'usinage et des configurations cinématiques d'un système de production reconfigurable. PhD thesis, Arts et Métiers ParisTech

Bejlegaard M, ElMaraghy W, Brunoe TD, Andersen AL, Nielsen K (2018) Methodology for reconfigurable fixture architecture design. CIRP J Manuf Sci Technol 23:172–186

Benderbal HH, Dahane M, Benyoucef L (2018) Modularity assessment in reconfigurable manufacturing system (RMS) design: an archived multi-objective simulated annealing-based approach. Int J Adv Manuf Technol 94(1–4):729–749

Benderbal HH, Dahane M, Benyoucef L (2015) A new robustness index for machines selection in reconfigurable manufacturing system. In: 2015 International conference on industrial engineering and systems management (IESM), pp 1019–1026. IEEE

Bensmaine A, Dahane M, Benyoucef L (2012) Process plan generation in reconfigurable manufacturing systems using AMOSA and TOPSIS. IFAC Proc Vol 45(6):560–565

Besharati-Foumani H, Lohtander M, Varis J (2019) Intelligent process planning for smart manufacturing systems: a state-of-the-art review. Procedia Manuf 38:156–162

Bi ZM, Lang SY, Shen W, Wang L (2008) Reconfigurable manufacturing systems: the state of the art. Int J Prod Res 46(4):967–992

Bortolini M, Galizia FG, Mora C (2018) Reconfigurable manufacturing systems: literature review and research trend. J Manuf Syst 49:93–106

Brahimi N, Dolgui A, Gurevsky E, Yelles-Chaouche AR (2019) A literature review of optimization problems for reconfigurable manufacturing systems. IFACPapersOnLine 52(13):433–438

Chryssolouris G (2006) Process planning. Manufacturing systems: theory and practice pp 281–328

Dahane M, Benyoucef L (2016) An adapted NSGA-II algorithm for a reconfigurable manufacturing system (RMS) design under machines reliability constraints. In: Metaheuristics for production systems, pp 109–130. Springer

Dashchenko AI (2006) Reconfigurable manufacturing systems and transformable factories. Springer

Deja M, Siemiatkowski MS (2013) Feature-based generation of machining process plans for optimised parts manufacture. J Intell Manuf 24(4):831–846

Dhupia J, Powalka B, Katz R, Ulsoy AG (2007) Dynamics of the arch-type reconfigurable machine tool. Int J Mach Tools Manuf 47(2):326–334

Dźwigoł H (2021) The uncertainty factor in the market economic system: the microeconomic aspect of sustainable development. Virtual Econ 4(1):98–117

ElMaraghy HA (2005) Flexible and reconfigurable manufacturing systems paradigms. Int J Flex Manuf Syst 17(4):261–276

Farid AM (2017) Measures of reconfigurability and its key characteristics in intelligent manufacturing systems. J Intell Manuf 28(2):353–369

Gadalla M, Xue D (2017) Recent advances in research on reconfigurable machine tools: a literature review. Int J Prod Res 55(5):1440–1454

Goyal KK, Jain PK (2016) Design of reconfigurable flow lines using MOPSO and maximum deviation theory. Int J Adv Manuf Technol 84(5):1587–1600

Goyal KK, Jain P, Jain M (2012) Optimal configuration selection for reconfigurable manufacturing system using NSGA II and TOPSIS. Int J Prod Res 50(15):4175–4191

Gu X, Koren Y (2022) Mass-individualisation–the twenty first century manufacturing paradigm. Int J Prod Res 1–16

Halevi G, Weill R (1994) Principles of process planning: a logical approach. Springer Science & Business Media

Jalonen H (2012) The uncertainty of innovation: a systematic review of the literature. J Manag Res 4(1):1

James T (2012) Smart factories. Eng & Technol 7(6):64–67

Katz R (2007) Design principles of reconfigurable machines. Int J Adv Manuf Technol 34(5):430–439

Khan AS, Homri L, Dantan JY, Siadat A (2022) An analysis of the theoretical and implementation aspects of process planning in a reconfigurable manufacturing system. Int J Adv Manuf Technol 1–32

Khanna K, Arneja GP, Kumar R (2014) Group technology in design of manufacturing systems-a review. In: Proceedings of the 5th international 26th All India manufacturing technology, design and research conference, pp 1–6

Koekemoer M, Gorlach I (2018) Development of a reconfigurable pallet system for a robotic cell. In: MATEC web of conferences, vol 210, p 02003. EDP Sciences

Koren Y (2010) The global manufacturing revolution: product-process-business integration and reconfigurable systems, vol 80. Wiley & Sons

Koren Y (2013) The rapid responsiveness of RMS. Int J Prod Res 51(23–24):6817–6827

Koren Y, Heisel U, Jovane F, Moriwaki T, Pritschow G, Ulsoy G, Van Brussel H (1999) Reconfigurable manufacturing systems. CIRP Ann 48(2):527–540

Koren Y, Gu X, Guo W (2018) Reconfigurable manufacturing systems: principles, design, and future trends. Front Mech Eng 13(2):121–136

Koren Y, Katz R (2003) Reconfigurable apparatus and method for inspection during a manufacturing process. US Patent 6,567,162

Koren Y (2006) General RMS characteristics. Comparison with dedicated and flexible systems. In: Reconfigurable manufacturing systems and transformable factories, pp 27–45. Springer

Koren Y (2020) The emergence of reconfigurable manufacturing systems (RMSS). In: Reconfigurable manufacturing systems: from design to implementation, pp 1–9. Springer

Kumar SL (2017) State of the art-intense review on artificial intelligence systems application in process planning and manufacturing. Eng Appl Artif Intell 65:294–329

Kumar G, Goyal KK, Batra NK, Rani D (2021) Single part reconfigurable flow line design using fuzzy best worst method. OPSEARCH 1–29

Landers RG, Min BK, Koren Y (2001) Reconfigurable machine tools. CIRP Ann 50(1):269–274

Liang CC (2013) Smart inventory management system of food-processing-and-distribution industry. Procedia Comput Sci 17:373–378

Ma Y, Du G, Jiao RJ (2020) Optimal crowdsourcing contracting for reconfigurable process planning in open manufacturing: a bilevel coordinated optimization approach. Int J Prod Econ 228:107884

Mei J, Zhang HC, Oldham WJ (1995) A neural network approach for datum selection in computer-aided process planning. Comput Ind 27(1):53–64
Moeuf A, Lamouri S, Pellerin R, Eburdy R, Tamayo S (2017) Industry 4.0 and the SME: a technology-focused review of the empirical literature. In: 7th International conference on industrial engineering and systems management IESM
Mohamad N, Rahman AA, Mohamad BB, Rahman M, Jafar F, Muhamad M, Fauadi MM (2018) Architecture of reconfigurable conveyor system in manufacturing system. J Adv Manuf Technol (JAMT) 12(1(2)):117–128
Molnar MF, Fernandez R, Huff B (2006) Development of a standard reconfigurable assembly cell for agile manufacturing. In: Reconfigurable manufacturing systems and transformable factories, pp 273–294. Springer
Morrar R, Arman H, Mousa S (2017) The fourth industrial revolution (Industry 4.0): a social innovation perspective. Technol Innov Manag Rev 7(11):12–20
Mubarok K (2020) Redefining Industry 4.0 and its enabling technologies. In: Journal of physics: conference series, vol 1569, p 032025. IOP Publishing
Niebel BW (1965) Mechanized process selection for planning new designs. ASME paper 737
Padayachee J, Bright G, Masekamela I (2009) Modular reconfigurable machine tools: design, control and evaluation. S Afr J Ind Eng 20(2):127–143
Pertuze JA, Reyes T, Vassolo RS, Olivares N (2019) Political uncertainty and innovation: the relative effects of national leaders' education levels and regime systems on firm-level patent applications. Res Policy 48(9):103808
Petrović M, Mitić M, Vuković N, Miljković Z (2016) Chaotic particle swarm optimization algorithm for flexible process planning. Int J Adv Manuf Technol 85(9):2535–2555
Prasad D, Jayswal S (2018) Scheduling of products for reconfiguration effort in reconfigurable manufacturing system. Mater Today Proc 5(2):4167–4174
Rossit DA, Tohmé F, Frutos M (2019) Industry 4.0: smart scheduling. Int J Prod Res 57(12):3802–3813
Rüßmann M, Lorenz M, Gerbert P, Waldner M, Justus J, Engel P, Harnisch M (2015) Industry 4.0: the future of productivity and growth in manufacturing industries. Boston Consult Group 9(1):54–89
Sautter B (2021) Shaping digital ecosystems for sustainable production: assessing the policy impact of the 2030 vision for Industrie 4.0. Sustainability 13(22):12596
Schwab K (2017) The fourth industrial revolution. Currency
Shao XF, Liu W, Li Y, Chaudhry HR, Yue XG (2021) Multistage implementation framework for smart supply chain management under Industry 4.0. Technol Forecast Soc Change 162:120354
Singh A, Asjad M, Gupta P (2019) Reconfigurable machine tools: a perspective. Life Cycle Reliab Saf Eng 8(4):365–376
Son H, Choi HJ, Park HW (2010) Design and dynamic analysis of an arch-type desktop reconfigurable machine. Int J Mach Tools Manuf 50(6):575–584
Tolio T, Ceglarek D, ElMaraghy H, Fischer A, Hu S, Laperrière L, Newman ST, Váncza J (2010) Species—co-evolution of products, processes and production systems. CIRP Ann 59(2):672–693
Tseng MM, Jiao J (2001) Mass customization. Handb Ind Eng 3:684–709
Wang Y, Zhang G, Han L (2019) A methodology of setting module groups for the design of reconfigurable machine tools. Int J Adv Manuf Technol 104(5):2133–2147
Wu W, Huang Z, Zeng J, Fan K (2021) A fast decision-making method for process planning with dynamic machining resources via deep reinforcement learning. J Manuf Syst 58:392–411

Xia Q, Etienne A, Jy D, Siadat A (2018) Reconfigurable machining process planning for part variety in new manufacturing paradigms: definitions, models and framework. Comput Ind Eng 115:206–219

Yelles-Chaouche AR, Gurevsky E, Brahimi N, Dolgui A (2021) Reconfigurable manufacturing systems from an optimisation perspective: a focused review of literature. Int J Prod Res 59(21):6400–6418

Yip-Hoi D (2017) Computer-aided process planning for machining. In: The mechanical systems design handbook, pp 11–37. CRC Press

Yusof Y, Latif K (2014) Survey on computer-aided process planning. Int J Adv Manuf Technol 75(1–4):77–89

Reconfigurable Supply Chain Network Design: Leveraging RMS Insights

Shokraneh K. Moghaddam and Angela D. Traill

Abstract This chapter presents a comparative analysis of Reconfigurable Manufacturing Systems (RMS) and Reconfigurable Supply Chains (RSC), examining their shared principles, distinct features, and the potential to transfer RMS insights to RSC development. RMS enables rapid adaptation to market fluctuations through modularity, scalability, and flexibility, revolutionizing how manufacturing systems respond to dynamic demands. Similarly, RSC aims to achieve agility across an interconnected network of suppliers, manufacturers, and distributors. The chapter explores how the foundational concepts of RMS—such as modular design, real-time decision-making, and scalable configurations—can inform and enhance the design of resilient supply chains capable of withstanding disruptions and demand volatility. The shift from RMS to RSC involves unique challenges including the complexity of coordinating multiple stakeholders, ensuring data interoperability, achieving end-to-end visibility, and balancing reconfiguration costs with performance benefits that are also addressed in the end.

1 Introduction

Reconfigurable Manufacturing Systems (RMS) are a modern approach to manufacturing designed to adapt quickly and efficiently to changing production needs. In contrast to dedicated manufacturing systems (DMS) that are mainly focused on mass production, or flexible manufacturing systems (FMS) that are dedicated to producing a wide range of high-quality products, an RMS can offer production in high volumes, customized flexibility, and the required responsiveness for adapting to unpredictable market changes (Koren et al. 2018; Moghaddam et al. 2018, 2020; Zhu et al. 2022;

S. K. Moghaddam (✉) · A. D. Traill
Department of Engineering, School of Physics, Engineering, and Computer Science, University of Hertfordshire, Hatfield, UK
e-mail: s.khashkhashimoghadam@herts.ac.uk

A. D. Traill
e-mail: a.traill@herts.ac.uk

S. Huang et al. (eds.), *Design and Operation of Smart Reconfigurable Manufacturing Systems in Industry 4.0/5.0*, Springer Series in Advanced Manufacturing,
https://doi.org/10.1007/978-3-032-00284-6_10

Huang et al. 2024; Saffar et al. 2025). This means it can be easily reconfigured to produce different products or to scale up and down in response to demand fluctuations. In other words, an RMS provides the functionality and capacity that is needed, when it is needed. Thus, a given RMS configuration can be dedicated or flexible, or in between, and can change as required.

The idea of configuration and reconfiguration has also been applied to how complex systems, like extended enterprise systems and supply chains, adapt and change. Thereby, a Reconfigurable Supply Chain (RSC) is defined as a network designed to be cost-efficient, responsive, sustainable, and resilient. RSC must be increasingly data-driven and dynamically adaptable, capable of quickly adjusting supply, production capacities, and functionality by rearranging, reallocating, or changing its components in response to sudden changes (Dolgui et al. 2020). The design of RSC networks is a critical area of study as it directly impacts the ability of organizations to navigate uncertainty and remain competitive.

In RMS paradigm, the key enablers for reconfiguration are the core characteristics that were defined by Koren et al. (1999) namely: Modularity, Integrability, Customization, Scalability, Convertibility, and Diagnosibility. Extensive research has been conducted on RMS configuration design, examining how each of the core characteristics impacts efficiency, lead time, cost, and other key performance indicators within a single manufacturing unit (Koren et al. 2018; Yelles-Chaouche et al. 2021; Bortolini et al. 2018). However, while these characteristics effectively capture the concept of reconfiguration, they have not been properly addressed in the context of an RSC. Consequently, their effectiveness within RSC network design has not yet been thoroughly explored.

In this article, we define the fundamental characteristics of RSC and explore key design and optimization parameters within both RSC and RMS. By highlighting the parallels between these two domains, we demonstrate how researchers can leverage some of these similarities to enhance the development of mathematical models and optimization strategies in RSC.

2 Core Characteristics of RMS and RSC

According to Koren et al. (2018), the following core characteristics are the major enablers for RMS adjustable structure. In what follows, based on the reviewed literature, we have also proposed definitions for these characteristics in the RSC context.

2.1 Modularity

This feature is the foundation of flexibility in an RMS. By designing the system with modular components, each module can function independently or in combination with others. This allows for the quick rearrangement or replacement of units to optimize the production process for different tasks or products. Modular designs reduce downtime and enhance the ability to adapt to new products or changes in production requirements.

Regarding RSC, past studies emphasized the importance of modularity in the reconfiguration of supply chains, arguing that modular systems allow for easier adaptation by breaking down the supply chain into manageable, interchangeable units (Zidi et al. 2021b). Based on our observations it can be defined as the ability to break down the supply chain into distinct, interchangeable components or segments. These could include suppliers, logistics providers, distribution centers, and even production facilities. Each component can be independently adjusted, replaced, or rearranged to optimize the supply chain for different products, markets, or conditions, enabling quick reconfiguration without overhauling the entire system.

2.2 *Integrability*

RMS relies on the seamless interaction between different modules. Integrability ensures that new modules or processes can be added to the system without disrupting existing operations. It involves standardized interfaces and communication protocols, which allow the various units to work together efficiently and share data in real-time, ensuring smooth and coordinated operations across the system.

Based on the past research, integrability in an RSC focuses on the seamless connection and communication between the various components of the supply chain (Zidi et al. 2021a, b; Dolgui et al. 2020). This means that different suppliers, manufacturers, logistics providers, and retailers can be integrated into the supply chain quickly and efficiently. Effective integrability ensures that these components work together smoothly, sharing information in real-time, which is crucial for coordinating activities and responding rapidly to changes in demand or supply.

2.3 *Customization*

The ability to produce a variety of products or variations within a product family is essential in today's market, where customers demand personalized products. RMS supports customization by allowing different product variants to be manufactured on the same system with minimal changes. This flexibility enables manufacturers to meet diverse customer needs while maintaining efficiency and cost-effectiveness.

According to Zidi et al. (2022b), customization in RSC is crucial in maintaining competitiveness, especially as markets demand increasingly personalized products. This charachteristic can be measured through various performance metrics such as the degree of product flexibility, the time added for value creation, throughput rate, and the number of customizable features available. Moreover, the level of customization is often evaluated by looking at customer satisfaction, the efficiency of new product development, and sustainability in the production process.

Hence, customization within an RSC can be defined as the capability to tailor the supply chain operations to meet the specific needs of different markets, customers, or product lines. This could mean adjusting the supply chain to handle different order sizes, delivery schedules, or packaging requirements, allowing for a more personalized approach to customer service and product delivery, which is increasingly important in today's consumer-driven market.

2.4 *Convertability*

Convertibility allows the RMS to switch between different products or product families with minimal downtime. This is achieved by designing machines and processes that can be easily reconfigured or adjusted to handle different tasks. Convertibility is key to responding to market changes or customer demands without the need for significant new investments or lengthy retooling processes.

Convertibility in an RSC refers to the ability to switch between different supply chain configurations or strategies with minimal effort and time (Zidi et al. 2022b). This might involve changing suppliers, altering transportation routes, shifting production locations, or adapting distribution strategies in response to shifts in market demand, cost structures, or external disruptions such as natural disasters or geopolitical events.

2.5 *Scalability*

As market demands fluctuate, the ability to scale production up or down is a significant advantage. Scalability in an RMS is achieved by easily adding or removing resources, such as machines, labor, or materials. This characteristic allows manufacturers to adjust their capacity to meet changing demand levels without major disruptions or overhauls. Scalability in an RSC refers to the ability to adjust the capacity of the supply chain up or down in response to fluctuations in demand. This could involve ramping up production, adding new suppliers, increasing logistics capacity, or conversely, scaling down operations during periods of low demand.

In RSC, scalability is typically measured by how well the supply chain can maintain its performance (throughput) while responding to increasing or varying workloads. This involves factors such as latency, throughput capacity, and the supply chain's ability to achieve performance objectives in dynamic and uncertain environments (Zidi et al. 2022a). Scalability ensures that the supply chain can maintain efficiency and cost-effectiveness regardless of the volume of goods being processed.

2.6 Diagnosability

This characteristic is crucial for maintaining high levels of productivity and quality. Diagnosability means the system can continuously monitor its performance, identify deviations or issues, and diagnose the root causes of problems quickly. This capability minimizes downtime and reduces the likelihood of defects reaching the customer by enabling fast corrective actions.

Studies such as those by Zidi et al. (2022a) and Dolgui et al. (2020) have highlightedthe significance of diagnosability in reconfigurable systems, linking it to other key reconfigurability characteristics such as modularity and convertibility. Diagnosability within a supply chain involves the system's ability to monitor itself continuously, detect inefficiencies, disruptions, or bottlenecks, and identify the root causes of these issues. By having a high level of diagnosability, a reconfigurable supply chain can quickly address problems such as delays, inventory shortages, or quality issues, thereby minimizing disruptions and maintaining optimal performance. A summary of all core characteristics of RSC that were defined in this section can be seen in Table 1.

3 Design of RMS Versus Network Design in Supply Chains

There has been extensive research regarding configuration design in RMS (Moghaddam et al. 2018, 2020). The research is heavily focused on balancing the objectives of cost minimization (material handling cost, module replacement cost, reconfiguration cost, etc.) and efficiency maximization, while addressing critical constraints such as capacity of resources and the precedence of operations. Supply chain network design and optimization usually looks for minimizing transportation costs and lead times, maximizing service levels, and enhancing overall efficiency.

Typical decision variables include location and capacity of critical infrastructure like warehouses, manufacturing facilities, distribution centers, etc. The major constraints are capacity limits for suppliers, manufacturers, and distributors. These limitations include restrictions on production capabilities, storage, and transportation capacities, all of which can cause bottlenecks if not properly accounted for in the planning process (Simchi-Levi et al. 1999). Other common constraints include market demand variability, supply disruptions, and uncertainties in lead times, all of which add complexity to the optimization process (Chopra and Meindl 2001). The following sections describe some of the key components in a supply chain and an RMS. While both systems involve numerous additional elements, this comparison will focus on a few essential components to highlight the most important similarities.

Table 1 Core characteristics of RMS Used in RSC context

Feature	RMS	RSC
Modularity	The division of operational functions into units that can be rearranged to fit different production setups	The division of supply chain into interchangeable components, such as suppliers or logistics providers, that can be easily rearranged to adapt to different products or market demands
Integrability	The capability to quickly and accurately integrate modules through standardized hardware and software interfaces	The capability to Seamlessly connect and coordinate different parts of the supply chain, including suppliers, manufacturers, and distribution centers, for smooth operations and real-time information sharing
Customization	The ability to offer system or machine flexibility within a specific part family, allowing for tailored flexibility within that family	Tailoring the supply chain to handle specific customer needs or product variations, providing flexibility in how goods are produced and delivered
Convertability	The capability to adapt the functionality of existing systems and machinery to meet new production requirements	Quickly reconfiguring the supply chain to meet new demands or conditions by adjusting strategies, switching suppliers, or altering logistics routes
Scalability	The ability to modify production capacity by either adding or removing resources or altering system components	The ability to modify the supply chain capacity by adding or removing resources or adjusting supply chain components
Diagnosability	The capacity to monitor product quality in real-time and quickly identify the root causes of any defects	The capacity of the supply chain to continuously monitor itself to detect issues, diagnose root causes of disruptions, and quickly address them to maintain efficiency

3.1 RMS Layout/Configuration Design and Optimization

RMS structure is designed to be flexible, allowing for the rapid adaptation of its structure to changes in product demand, product design, or production volume. In such a system, several key components and characteristics are typically visible (shown in Fig. 1):

- **Modular RMTs and Equipment**: Machines and equipment in an RMS are modular, meaning they can be easily reconfigured. These machines often have adjustable settings and capabilities to adapt to different tasks. Some machines can be retooled quickly to produce different components or products.
- **Interchangeable Modules**: Some sections of the layout may include interchangeable modules that can be swapped out to accommodate different operations, such as modules for milling, turning, or assembly. These modules can be customized to specific production requirements and scaled up or down.

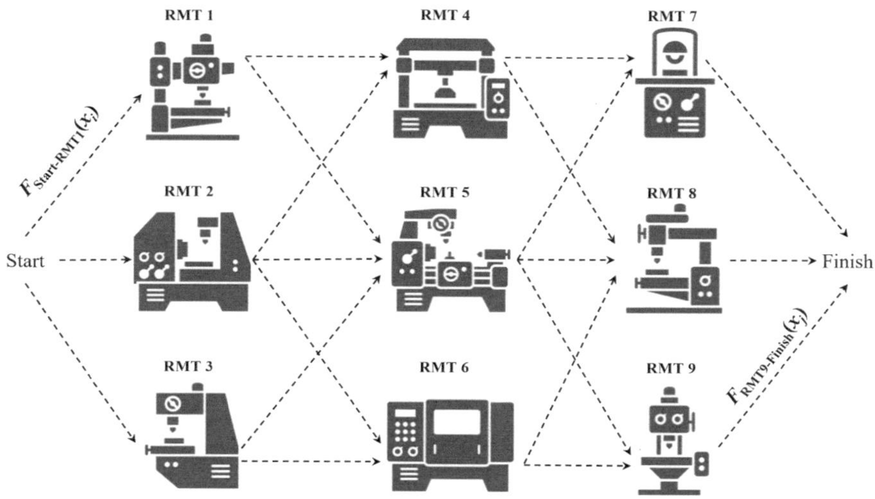

Fig. 1 Possible layout of an RMS

- **Automated Material Handling Systems**: Automated material handling systems like AGVs (Automated Guided Vehicles), conveyors, or robotic arms are common, enabling the efficient transport of materials between workstations or production cells. These systems are typically designed to adapt to different configurations of the layout.
- **Changeable System Layout**: The layout supports scalability, allowing for the addition or removal of production lines or machines as demand changes. The layout itself is designed to change over time, meaning that pathways for material flow, worker movement, and machine placement are often flexible.
- **Adjustable Flow**: This normally is the flow of work-in-progress (WIP) associated with a certain part which moves from one resource to other resources in the system. For example, in Fig. 1, $F_{Start-RMT1}(x_i)$ represents that x amount of WIP associated with part i is moving from Start point to RMT 1. The flow can be adjusted in RMS based on demand fluctuations.

3.2 *Supply Chain Network Design and Optimization*

In a typical supply chain network one can observe the interconnected flow of goods and information between different entities, each playing a key role inmoving products from raw materials to the end consumer (Fig. 2). The network is composed of several layers and functions that work together to ensure efficient product flow. This network is often visualized as a series of interconnected nodes (suppliers, manufacturers, warehouses, retailers) and links (transportation, information flows). The following are some of the key components:

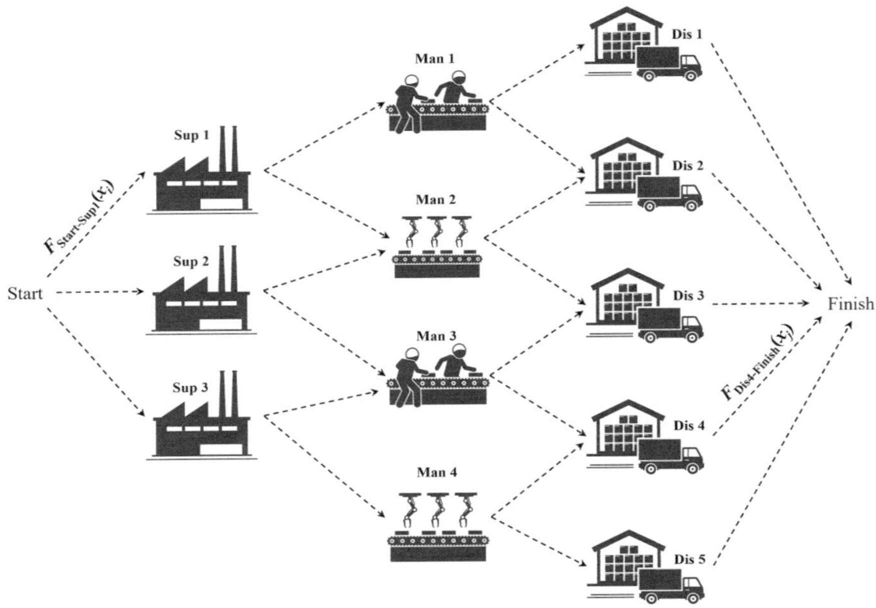

Fig. 2 Possible layout of a supply chain network

- **Suppliers**: Can provide raw material such as metals, plastics, textiles, chemicals, or components and they can also deliver parts and sub-assemblies that are integrated into final products, such as electronic parts, mechanical assemblies, etc. Suppliers are often categorized into tiers (tier 1, tier 2, etc.), representing how closely they work with manufacturers.
- **Manufacturing and Production Centers**: Convert raw materials and components into finished goods. This includes factories, assembly plants, and processing facilities. These facilities often use inputs from various suppliers and transform them into products through processes like assembly, machining, or packaging.
- **Warehouses and Distribution Centers**: Warehouses hold raw materials, work-in-progress, or finished goods at various stages of the supply chain. They are used to store products before they are needed in manufacturing or distribution. Distribution centers focus on efficiently receiving goods from manufacturers and distributing them to retailers or consumers. They are more focused on the rapid movement of goods.
- **Transportation and Logistics Providers**: Various forms of transportation can be used, such as trucks, ships, trains, and planes, which are used to move goods between different nodes (suppliers, manufacturers, warehouses, retailers).
- **Flow**: This can be the flow of materials from supply chain upstream to downstream and also information flow provided by software systems used to manage and integrate supply chain operations, such as procurement, production, logistics, and sales.

Clearly, an RMS can operate as a distinct yet integral component within a broader supply chain, as illustrated in Fig. 3. There are also noticeable similarities between the elements of an RMS and those in a traditional supply chain. Both systems consist of interdependent processes, resources, and operations that must be coordinated to ensure smooth production and delivery. For example, just as a supply chain comprises suppliers, manufacturers, distributors, and retailers, an RMS involves various modular components like machine tools, control systems, and production units that can be reconfigured or scaled up/down based on demand. Therefore, by applying the same terminologies used in an RMS to the supply chain context, we can define parallel components and processes in an RSC (as shown in Table 2). Table 2 also signifies the fact that many of the cost and time associated parameters can be matched simultaneously.

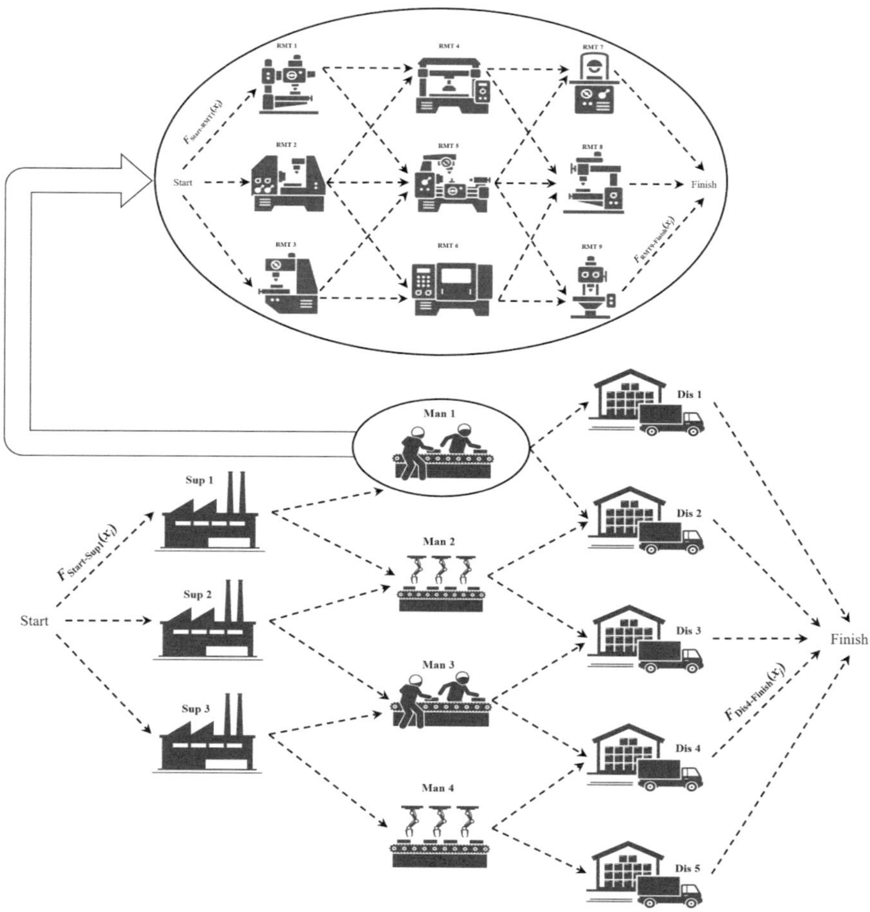

Fig. 3 An RMS embedded in a supply chain network

Table 2 Similarities between RSC and RMS

	System components	Related information	Associated costs	Associated time
RMS context	RMTs	– Different ***configurations*** – ***Functionality*** of each configuration – ***Capacity*** of each configuration	– Cost of purchase – Cost of changing functionality and/or capacity	– Time to perform a certain function – Time to change functionality and/or capacity
RSC context	Suppliers Manufacturers Distributors	– Different ***configurations*** – ***Functionality*** of each configuration – ***Capacity*** of each configuration	– Cost of making a contract – Cost of changing functionality and/or capacity	– Time to perform a certain function – Time to change functionality and/or capacity
RMS context	Part family	– Demand for different parts – ***Resources*** required for each part	– Production cost per part	– Production time per part
RSC context	Products	– Demand for different products – ***Resources*** required for each product	– Supply chain network costs per product	– Lead time per product
RMS context	Layout	– Possible installation locations – Distance between locations	– Material handling cost – Cost of relocating RMTs (if applicable)	– Material handling time – Time for relocating RMTs
RSC context	Network	– Possible geographical locations – Distance between locations	– Logistics costs – Cost of adding/removing suppliers, manufacturers, or distributors	– Logistics time required – Time required for adding/removing suppliers, manufacturers, or distributors

Definitions of *italic* words differ in RMS and RSC context. Specific definitions are provided in Table 3 for more clarity

While the terminologies and functional roles may overlap, the key difference lies in scale and focus. RMS is primarily focused on manufacturing processes and the reconfigurability of production lines, whereas a supply chain emphasizes the end-to-end movement of goods, including sourcing, production, and delivery to the final customer. In Table 3, major differences in definitions of similar terms that can be used in both contexts are further elaborated.

In our analysis, we have focused on some of the most critical components of the RSC and RMS, specifically suppliers, manufacturers, distributors, and RMTs. While we have limited the scope of our investigation to these key elements, this does not reduce the general applicability of our findings. Although we have chosen

Table 3 Definitions of terms in RSC and RMS contexts

Term	Context		Definition	Enablers
Configuration	RMS	RMT	The setup for each RMT consisting of fixed and auxiliary modules that can be added or removed to adjust production requirements	Modularity
		System	Specific arrangement and setup of the manufacturing system, including its hardware, software, and operational parameters, that can be adapted or altered to meet changing production requirements	Modularity Integrability
	RSC	Suppliers	Specific setup and organization of the supplier's operations, resources, and processes that enable them to adapt quickly and efficiently to changing demands, production requirements, and supply chain dynamics	Modularity Integrability
		Manufacturers	Specific setup and organization of the manufacturer's production systems, processes, resources, and relationships that enable the company to adapt rapidly to changing market conditions, customer demands, and supply chain dynamics	Modularity Integrability
		Distributors	Strategic setup and organization of the distributor's logistics, inventory management, transportation, and customer service operations that enable the company to adapt quickly and efficiently to changing market conditions, customer demands, and supply chain dynamics	Modularity Integrability
		Network	Specific arrangement and organization of supply chain components, including suppliers, manufacturers, distribution centers, and logistics providers, that can be dynamically adjusted to respond to changing market conditions, demand patterns, or disruptions	Modularity Integrability
Functionality	RMS	RMT	Performing different operations (e.g., drilling, milling, or cutting) with different levels of complexity for different products	Convertibility

(continued)

Table 3 (continued)

Term	Context		Definition	Enablers
	RSC	Suppliers	Performing a wide range of operations (e.g., providing raw materials, components, finished goods, or services) for different products	Convertibility
		Manufacturers	Performing a wide range of operations to convert raw materials or components into finished goods through various production processes	Convertibility
		Distributors	Performing a wide range of activities (e.g., logistics, inventory management, and customer service) as intermediaries between manufacturers and retailers or end-users to ensure that products are available in the right place, at the right time, and in the right quantity	Convertibility
Capacity	RMS	RMT	Maximum rate at which different operations (e.g., drilling, milling, or cutting) are performed for different parts in a certain period	Scalability
		System	Maximum production volume for different products belonging to a part family in a certain period	Scalability
	RSC	Suppliers	Maximum units of raw materials, components, finished goods, etc. which can be supplied for different products in a certain period	Scalability
		Manufacturers	Maximum production volume for different products in a certain period	Scalability
		Distributors	Maximum number of different products that can be distributed in a certain period	Scalability
		Network	Maximum number of different products that can be supplied, produced, and distributed in a certain period	Scalability
Resources	RMS		RMTs that are used to produce different parts	Customization
	RSC	Suppliers	All tangible (e.g., raw material, manufacturing equipment, machinery) and intangible (e.g., knowledge, relationships) assets that are used to supply goods efficiently	Customization

(continued)

Table 3 (continued)

Term	Context		Definition	Enablers
		Manufacturers	All tangible (e.g., basic material, manufacturing equipment, machinery and tooling, technology and software systems) and intangible (e.g., intellectual properties, knowledge, relationships) assets that are used to produce goods efficiently	Customization
		Distributors	All logistics, warehousing, inventory management, customer service, and other assets used to ensure goods reach their final destinations efficiently	Customization

to concentrate on these particular components for the sake of clarity and simplicity, other equally important elements of the supply chain—such as retailers, customers, logistics and transportation systems—can be analyzed using the same principles and comparative framework.

4 Optimization Opportunities for RSC: An Example

Based on the proposed terminologies, many of the mathematical models proposed earlier for configuration design in RMS can be used for addressing network design decisions in RSC. Here we propose a very simple example from Rezaee and Moghaddam (2025) where the addressed problem is designing an RMS configuration for a scalable production system that can manufacture a variety of products within the same part family. To meet products demand at the lowest possible cost, the initial RMS configuration must adapt throughout the product life-cycle as demand rates for each product fluctuate. A predictive method is proposed to determine the system configuration for each production phase, based on the predicted demand.

Based on the definitions provided in Tables 2 and 3, parameters and decision variables defined for the proposed integer linear programming (ILP) model (shown in (1)) can be adjusted to address the problem of network design in RSC as shown in Table 4. The proposed mathematical model can be solved for both realms given the related parameters and data are collected accordingly.

$$\min \sum_{i \in I} \sum_{j \in J} \sum_{k \in K} \sum_{t=T} C_{ij} w_{ijkt} + \sum_{i \in 1} \sum_{j \in J} \sum_{j' \in J} \sum_{k \in K} \sum_{t=T} B_{ijj'} z_{ijj'kt} \quad (1)$$

Subject to:

Table 4 Definitions of parameters and decision variables in RMS and RSC contexts

		Definition in RMS context	Definition in RSC context
Parameters	$\boldsymbol{T}$	Number of periods	Number of periods
	$\boldsymbol{K}$	Total number of functions	Total number of functions
	$\boldsymbol{I}$	Total number of available RMTs	Total number of available resources
	$\boldsymbol{J}$	Number of possible RMT configurations	Number of possible configurations for resources
	$\boldsymbol{D_{kt}}$	Demand rate for the kth function in period t	Demand rate for the kth function in period t
	$\boldsymbol{C_{ij}}$	Cost of purchasing the ith RMT in its jth configuration	Cost of selecting the ith resource in its jth configuration
	$\boldsymbol{P_{ijk}}$	Capacity of the ith RMT in its jth configuration for performing kth function	Capacity of the ith resource in its jth configuration for performing kth function
	$\boldsymbol{B_{ijj'}}$	Cost of transforming RMT type i from its jth configuration to j'th configuration	Cost of transforming resource type i from its jth configuration to j'th configuration
Decision variables	$\boldsymbol{w_{ijkt}}$	An integer decision variable that determines the number of RMTs of type i in their jth configuration performing kth function purchased at the beginning of period t	An integer decision variable that determines the number of resources of type i in their jth configuration performing kth function selected at the beginning of period t
	$\boldsymbol{z_{ijj'kt}}$	An integer decision variable that determines the number of RMTs of type i in their jth configuration that are transformed into j'th configuration to perform kth function during period t	An integer decision variable that determines the number of resources of type i in their jth configuration that are transformed into j'th configuration to perform kth function during period t

$$\sum_{i\in I}\sum_{j\in J} P_{ijk} w_{ijkt} + \sum_{i\in I}\sum_{j\in J}\sum_{j'\in J} P_{ijk} z_{ijj'kt} \geq D_{kt} \quad \forall k, t \tag{2}$$

$$\sum_{k\in K} w_{ijk(t-1)} + \sum_{j'\in J}\sum_{k\in K} z_{ij'jk(t-1)} = \sum_{j'\in J}\sum_{k\in K} z_{ijj'kt} \quad \forall i, j, t > 1 \tag{3}$$

$$w_{ijkt}, z_{ijj'kt} \in \text{int} \quad \forall i, j, j', k, t \tag{4}$$

In the RMS domain, the primary objective of the optimization model, represented by Eq. (1), is to minimize the combined costs associated with purchasing new RMTs and the reconfiguration expenses incurred across all production periods. Constraint (2) ensures that the projected demands for products are fully satisfied during each production period. This demand can be met through two main avenues: either by acquiring additional RMTs or by reconfiguring the existing equipment. Constraint (3) establishes balance during reconfiguration periods by ensuring that the number and types of RMTs remain consistent. Specifically, the sum of RMT purchased, reconfigured, or left unchanged in one production period must equal the total number of RMTs available at the start of the subsequent period. Lastly, constraint (4) defines the allowable domains for the decision variables, ensuring they adhere to logical and operational bounds.

For the RSC domain, based on the definitions provided in Table 4 the objective of (1) would be minimizing costs of selecting new resources as well as reconfiguration costs in all production periods. Constraint (2) ensures that predicted demands are satisfied in all production periods. Products' demand can be met through either selecting new resources or resource reconfigurations. Constraint (3) guarantees equal number and types of resources during reconfiguration periods i.e., the number of resources with certain configurations that are selected, reconfigured, or remained unchanged during one production period must be equal to the number of resources at hand in the next period. Similarly, domains of variables are defined in constraint (4).

5 Key Barriers to Reconfigurable Supply Chain

Implementing an RSC paradigm presents numerous challenges, each rooted in the complex nature of modern supply chain networks. One of the primary hurdles is technological integration. RSCs rely heavily on advanced tools such as IoT, digital twins, and AI to enable real-time monitoring and reconfiguration. However, incorporating these technologies into legacy systems often leads to compatibility issues, lack of standardization, and high implementation costs. For many organizations, particularly small and medium enterprises, the expense and technical expertise required for such upgrades can be prohibitive (Gunasekaran et al. 2019).

Cost implications are another significant barrier. The transition to an RSC paradigm requires substantial investment in reconfigurable equipment, software, and employee training programs. While these investments promise long-term benefits, the high upfront costs can deter organizations from adopting the model. This is particularly relevant for industries that operate on thin profit margins or face market volatility, making long-term commitments risky.

The complexity of decision-making in RSCs adds another layer of difficulty. Dynamic reconfiguration involves balancing multiple objectives, such as cost, speed, and resource utilization. The lack of robust decision-support systems capable of handling this complexity often results in suboptimal decisions. Wang et al. (2021) emphasize the need for multi-objective optimization frameworks to address this challenge and enhance adaptability.

Data management and cybersecurity are critical concerns in the RSC paradigm. Reliable reconfiguration depends on real-time, accurate data. Ensuring data integrity and managing vast amounts of information while protecting against cybersecurity threats are daunting tasks. Breaches or inaccuracies can disrupt supply chain operations and erode trust among stakeholders (Christopher and Peck 2004).

Finally, an RSC is one that responds readily to the changing needs of suppliers and manufacturers, market dynamics and disruption to the flow of materials, goods, data and finance. Another way of describing the RSC is as an 'adaptive' system. To be adaptive to change, a high degree of connectivity is required: to stakeholders in the end-to-end supply chain; to assets (people, facilities, and equipment) working in the supply chain; to data regarding those assets and the factors which affect them, including infrastructure conditions. An intricate network of connections and interactions is required to then coordinate and synchronize supplies, production and logistics with the ultimate delivery requirements of the customer (including reverse logistics to cater for a growing circular economy).

The premise that an RSC requires connectivity between established networks of stakeholders throughout a supply chain, presents us with another quite important barrier. One could cite three key elements required to achieve the level of connectivity needed for a supply chain to be truly reconfigurable:

- Collaborative Business Models that support collaborative, networked operations.
- Big data collection, management and sharing.
- Computing power to enable fast analytics for rapid, adaptive and agile decision making.

To suppose that these are separate elements, to deal with one-by-one, would be a mistake: anyone that has tried even the simplest of collaborations knows this. A holistic approach is required which deals with all these elements in parallel.

At the heart of the matter are people—decision-makers, managers, leaders and teams of 'doers' and the trust which provides the glue between them. If they do not trust those they would need to connect with or to collaborate with, or they are driven by conflicting visions, aims and objectives, there is little chance of any supply chain being rapidly and successfully reconfigured to changing market and business circumstances: Required data will not be collected, managed or shared openly; acceptance of automated decision-making or options put before them to choose from are less likely to be followed; commitment of required resources, time and investment in the next generation of software and hardware, algorithms and analytics tools, are unlikely to manifest. Aside from all this, the world is still arguably some years away from the quantum computing capabilities required to achieve the level of 'data-crunching' for the most complex systems analytics in support of automated, near or real-time decision-making that a truly adaptive or RSC would need.

Such challenges have been the focus of the Alliance for Logistics Integration and Coordination in Europe (ALICE), European Technology Platform, for over twenty years. This research and development network adopted the concept of the Physical Internet (PI), born from research of Montreuil (2011). The ALICE network devised a research roadmap, aiming at the EU's Net-Zero logistics policy agenda by 2050. The use of a PI is seen as a strategically critical in the achievement of net-zero. Nevertheless, the PI still appears some way off from any commercially significant application; this is in spite of some impressive results from small-scale trials. Technology still needs to prove itself; industry is unlikely to act until this happens and the business-case for investment is proven. For it to be proven, leadership is required, with public and private collaboration to de-risk development prior to scale economies to emerge and costs to come down to acceptable levels for the bulk of industry to engage.

The PI is, in itself, reliant on the modularity of logistics services and functions. It works on the premises that services, functions and equipment are interoperable, scalable, abundant to easily achieve critical mass, and sufficiently flexible and agile to suit all demands placed on the network to deliver what the customer requires, when they want it, at the quality and price they want and with adequate profit equitably shared among the network of participants. The network must be extensive and inclusive. Data must be made available to the system as a whole in order to synchronize the availability of the assets 24/7. When customers, products or markets change, the network will supply the right type of assets available within the network's pool of assets, to deliver as and when, where and at an optimal cost to price ratio, acceptable to participants and the customers. Whatever the size of shipment, value, volume, shape, frequency, origin, or destination, the PI-enabled logistics network will be able to customize delivery to suit the customer on a case-by-case basis.

Whatever the disturbance to the supply chain, an optimal logistics solution will be available, through instant diagnostics. The data and the trust required of participants and by participants, asset owners, operators, infrastructure managers, even policy makers and compliance enforcers, will be huge. Cybersecurity will be of paramount importance. Blockchain systems incorporated within the PI-enabled network would arguably provide a degree of impenetrability to systems holding valuable and confidential data, preventing unauthorized access to the systems by virtue of a federated network of servers and ERP and customer management systems.

Trust, remains the biggest barrier, however, to the implementation of a PI-enabled logistics network at the scale envisaged here. A model framework was proposed to address the eight areas that drive greater trust within collaboration networks (Sun et al. (2024)):

- **Adoption**: Achieving critical mass through common policies and technologies.
- **Altruism**: Putting collective benefits above individual interests in collaborative efforts.
- **Compliance**: Following agree norms and standards to ensure reliability and predictability.
- **Confidentiality**: Ensuring data is accessible only to authorized parties and protected from exposure.
- **Control**: Exercising authority over data and cargo to align decisions with organizational preferences.
- **Interest**: Expected individual or organizational benefits from collaboration.
- **Reputation**: Reliability based on past behavior and adherence to ethical standards.
- **Transparency**: Availability of clear information and traceability to promote openness and accountability.

Without the modularity of logistics, this form of collaborative model might fail to materialize. More 'relational' and 'captive' forms of collaboration, seeking hierarchical governance to increase supply chain value, would act as further barriers to

RSCs. These latter models base themselves on a lead firm extending far greater influence over its suppliers, such that they exert control over much of what those suppliers do, how they do it and of course, the resulting price of what they offer. Suppliers become much more integrated into and, in the most extreme cases, captives of the lead firms. Switching to new suppliers or into new markets with new customers becomes much more difficult and costly. The more inter-connected to lead firms a supplier becomes, the less resilient they become to disruption events: they cannot readily adopt multi-supplier strategies or new suppliers—a strategy recognized among those proposed by business-government organizations like the World Economic Forum.

In conclusion, while RSCs offer immense potential for enhancing flexibility, resilience, and sustainability in supply chain networks, their successful implementation demands careful attention to these challenges. Addressing these barriers through critical to realizing the full benefits of this paradigm.

References

Bortolini M, Galizia FG, Mora C (2018) Reconfigurable manufacturing systems: literature review and research trend. J Manuf Syst 49:93–106

Chopra S, Meindl P (2001) Strategy, planning, and operation. Supply Chain Manag 15(5):71–85

Christopher M, Peck H (2004) Building the resilient supply chain. Int J Logist Manag 15(2):1–13

Dolgui A, Ivanov D, Sokolov B (2020) Reconfigurable supply chain: the x-network. Int J Prod Res 58(13):4138–4163

Gunasekaran A, Subramanian N, Papadopoulos T (2019) Information technology for competitive advantage within logistics and supply chains: a review. Transp Res Part E Logist Transp Rev 99:14–33

Huang S, Tan J, Lu Y, Moghaddam SK, Wang G, Yan Y (2024) A multiobjective joint optimisation method for simultaneous part family formation and configuration design in delayed reconfigurable manufacturing system (D-RMS). Int J Prod Res 62(1–2):92–109

Koren Y, Heisel U, Jovane F, Moriwaki T, Pritschow G, Ulsoy G, Van Brussel H (1999) Reconfigurable manufacturing systems. CIRP Ann 48(2):527–540

Koren Y, Gu X, Guo W (2018) Reconfigurable manufacturing systems: principles, design, and future trends. Front Mech Eng 13:121–136

Moghaddam SK, Houshmand M, Fatahi Valilai O (2018) Configuration design in scalable reconfigurable manufacturing systems (RMS); a case of single-product flow line (SPFL). Int J Prod Res 56(11):3932–3954

Moghaddam SK, Houshmand M, Saitou K, Fatahi Valilai O (2020) Configuration design of scalable reconfigurable manufacturing systems for part family. Int J Prod Res 58(10):2974–2996

Montreuil B (2011) Toward a physical internet: meeting the global logistics sustainability grand challenge. Logist Res 3:71–87

Rezaee P, Moghaddam SK (2025) Assessment of design approaches for reconfigurable manufacturing systems based on forecasted demand data. Comput & Ind Eng 110878

Saffar F, Moghaddam SK, Huang S (2025) Dynamic layout design for scalable reconfigurable manufacturing systems. Int J Prod Res 1–23. https://doi.org/10.1080/00207543.2025.2474215

Simchi-Levi D, Kaminsky P, Simchi-Levi E (1999) Designing and managing the supply chain: concepts, strategies, and cases. McGraw-Hill, New York

Sun S, Michiels P, Macharis C, Cant A, Van Bever D, Mommens K (2024) Unlocking the potential of the physical internet: a trust-enabling decentralized process sharing connector. In: Proceedings

of the IPIC 2024 conference papers and posters contributions proceedings, pp 29–31. Savannah, GA, USA
Wang Y, Ren L, Zhou W, Zhang H (2021) Multi-objective optimization in supply chains: applications and advancements. Int J Prod Econ 235:108120
Yelles-Chaouche AR, Gurevsky E, Brahimi N, Dolgui A (2021) Reconfigurable manufacturing systems from an optimisation perspective: a focused review of literature. Int J Prod Res 59(21):6400–6418
Zhu Q, Huang S, Wang G, Moghaddam SK, Lu Y, Yan Y (2022) Dynamic reconfiguration optimization of intelligent manufacturing system with humanrobot collaboration based on digital twin. J Manuf Syst 65:330–338
Zidi S, Hamani N, Kermad L (2022a) Antecedents and enablers of supply chain reconfigurability and their effects on performance. Int J Adv Manuf Technol 120(5):3027–3043
Zidi S, Hamani N, Kermad L (2022b) New metrics for measuring supply chain reconfigurability. J Intell Manuf 33(8):2371–2392
Zidi S, Hamani N, Kermad L (2021a) Classification of reconfigurability characteristics of supply chain. In: Proceedings of the changeable, agile, reconfigurable and virtual production conference and the world mass customization & personalization conference, pp 72–79. Springer
Zidi S, Hamani N, Kermad L (2021b) Modularity metric in reconfigurable supply chain. In: IFIP international conference on advances in production management systems, pp 455–464. Springer

MIX
Papier aus verantwortungsvollen Quellen
Paper from responsible sources
FSC® C105338

If you have any concerns about our products,
you can contact us on
ProductSafety@springernature.com

In case Publisher is established outside the EU,
the EU authorized representative is:
Springer Nature Customer Service Center GmbH
Europaplatz 3, 69115 Heidelberg, Germany

Printed by Libri Plureos GmbH
in Hamburg, Germany